Scaling in Soil Physics: Principles and Applications

Scaling in Soil Physics: Principles and Applications

Proceedings of a symposium sponsored by Division S-1 of the Soil Science Society of America in Las Vegas, NV, 18 Oct. 1989. The symposium was held in honor of E.E. Miller and R.D. Miller.

Editors

Daniel Hillel
David E. Elrick

Organizing Committee

David E. Elrick
Daniel Hillel, chair

Editor-in-Chief SSSA

D.E. Kissel

Managing Editor

Susan Ernst

SSSA Special Publication Number 25

**Soil Science Society of America, Inc.
Madison, Wisconsin, USA
1990**

Cover Design: Patricia J. Scullion

Soil Science Society of America, Inc.
677 South Segoe Road, Madison, WI 53711, USA

Library of Congress Cataloging-in-Publication Data
Scaling in soil physics, principles and applications: proceedings of a symposium/sponsored by Division S-1 of the Social Science Society of America in Las Vegas, NV, 18 Oct. 1989: editors, David E. Elrick, Daniel Hillel.
p. cm.—(SSSA special publication: no. 25)
Includes bibliographical references.
ISBN 0-89118-792-8
1. Soil physics—Congresses. 2. Fractals—Congresses.
1. Elrick, David E. (David Emerson), 1931- . II. Hillel, Daniel. III. Soil Science Society of America. Division S-1. IV. Series.
631.4′3—dc20

90-45642
CIP

This book is printed on acid-free paper.

Printed in the United States of America

CONTENTS

FOREWORD

The extreme variability of soils has made it very difficult to generalize our quantitative physical theories that are derived for ideal, homogenous systems. In hindsight, the concept of similitude seems both obvious and logical. Unfortunately, despite the current interest in geostatistical approaches to soil variability, the full power of the similarity concepts has yet to be realized in practice. The papers in this symposium should make this concept much more accessible to soil scientists, earth scientists, and engineers. Properly applied, similarity should bring the same insight and conceptual structure to soil science that dimensional analysis has brought to engineering and physics. The papers in this publication deserve careful study and analysis.

W.R. GARDNER, *president*
Soil Science Society of America

THE MILLERS

Edward E. Miller

Robert D. Miller

PREFACE

The theory of similitude and the attendant technique of scaling have long been used in applied physics to facilitate the analysis of varied problems. The principle is to formulate the relevant equations with the smallest possible number of variables, by clustering the variables and casting them into dimensionless form. The equations are thereby generalized and made applicable to any set of actual cases, provided the systems described are essentially similar (i.e., "scale models" of one another). The very exercise of scaling helps to reveal the fundamental relationships among operating variables.

It was only in 1955, however, that the concepts of scaling and similitude were introduced into soil physics by the brothers Ed and Bob Miller. In their twin papers, published that year in the *Soil Science Society of America Proceedings*, and in their subsequent paper published the following year in the *Journal of Applied Physics*, the Millers formulated the basic theory and defined the appropriate criteria for its application to surface tension-viscous flow phenomena in unsaturated porous media. Their seminal analysis provided new insights into the physical behavior of soil-water systems and has been applied ever since to the solution of many otherwise vexing problems, notably including the characterization of spatial variability, of hysteretic and unstable flow phenomena, and of mechanical stress distribution in unsaturated granular soils.

In a timely effort to acknowledge and highlight the landmark achievement of the Miller brothers, and to review subsequent developments in the area of similitude and scaling of soil systems, the Soil Physics Division of the Soil Science Society of America convened a special symposium as part of the Society's annual meetings in Las Vegas in October of 1989. This publication is a compendium of the invited papers presented at that symposium. As organizers of the symposium, we dedicate this volume to Ed and Bob Miller, with high regard for their inspired and inspiring contributions to soil physics. All who know the work of the Millers admire their outstanding professional qualities of originality, rigor, and insight. And all who have had the good fortune to associate with the Millers and learn from them directly cannot but feel an affection and a deep appreciation for their exemplary personal qualities of integrity, generosity, humility, friendliness, and—last but not least—perpetual good cheer.

Editors

DANIEL HILLEL
University of Massachusetts
Amherst, Massachusetts

DAVID E. ELRICK
University of Guelph
Guelph, Ontario, Canada

CONTRIBUTORS

R. J. Glass	Doctor, Sandia National Laboratory, Geoscience Analysis Division, Albuquerque, NM 87185
W. A. Jury	Professor of Soil Physics, Department of Soil and Environmental Sciences, University of California-Riverside, Riverside, CA 92521
Edward E. Miller	Professor Emeritus of Physics and Soil Science, Department of Physics, University of Wisconsin-Madison, Madison, WI 53706
R. D. Miller	Professor Emeritus of Soil Physics, Department of Soil, Crop, and Atmospheric Sciences, Cornell University, Ithaca, NY 14853
J.-Y. Parlange	Professor of Agricultural Engineering, Cornell University, Ithaca, NY 14853
P. A. C. Raats	Senior Research Scientist and Professor of Continuum Mechanics, Department of Mathematics, Wageningen Agricultural University; and Institute for Soil Fertility Research, 9750 RA Haren, the Netherlands
Victor A. Snyder	Associate Soil Scientist, Department of Agronomy and Soils, Agricultural Experiment Station, University of Puerto Rico, Rio Piedras, PR 00928
Garrison Sposito	Professor of Soil Physical Chemistry, Department of Soil Science, University of California-Berkeley, Berkeley, CA 94720
T. S. Steenhuis	Associate Professor, Department of Agricultural and Biological Engineering, Cornell University, Ithaca, NY 14853
Scott W. Tyler	Assistant Research Soil Scientist, Water Resources Center, Desert Research Institute, Reno, NV 89506
A. W. Warrick	Professor of Soil Physics, Soil and Water Science Department, University of Arizona, Tucson, AZ 85721
Stephen W. Wheatcraft	Associate Research Professor of Hydrogeology, Desert Research Institute, University of Nevada System, Reno, NV 89506
E. G. Youngs	Visiting Professor of Soil Physics, Silsoe College, Cranfield Institute of Technology, Silsoe, Bedford MK45 4DT, England

Conversion Factors for SI and non-SI Units

To convert Column 1 into Column 2, multiply by	Column 1 SI Unit	Column 2 non-SI Unit	To convert Column 2 into Column 1, multiply by
	Length		
0.621	kilometer, km (10^3 m)	mile, mi	1.609
1.094	meter, m	yard, yd	0.914
3.28	meter, m	foot, ft	0.304
1.0	micrometer, μm (10^{-6} m)	micron, μ	1.0
3.94×10^{-2}	millimeter, mm (10^{-3} m)	inch, in	25.4
10	nanometer, nm (10^{-9} m)	Angstrom, Å	0.1
	Area		
2.47	hectare, ha	acre	0.405
247	square kilometer, km^2 $(10^3\ m)^2$	acre	4.05×10^{-3}
0.386	square kilometer, km^2 $(10^3\ m)^2$	square mile, mi^2	2.590
2.47×10^{-4}	square meter, m^2	acre	4.05×10^{3}
10.76	square meter, m^2	square foot, ft^2	9.29×10^{-2}
1.55×10^{-3}	square millimeter, mm^2 $(10^{-3}\ m)^2$	square inch, in^2	645
	Volume		
9.73×10^{-3}	cubic meter, m^3	acre-inch	102.8
35.3	cubic meter, m^3	cubic foot, ft^3	2.83×10^{-2}
6.10×10^{4}	cubic meter, m^3	cubic inch, in^3	1.64×10^{-5}
2.84×10^{-2}	liter, L (10^{-3} m^3)	bushel, bu	35.24
1.057	liter, L (10^{-3} m^3)	quart (liquid), qt	0.946
3.53×10^{-2}	liter, L (10^{-3} m^3)	cubic foot, ft^3	28.3
0.265	liter, L (10^{-3} m^3)	gallon	3.78
33.78	liter, L (10^{-3} m^3)	ounce (fluid), oz	2.96×10^{-2}
2.11	liter, L (10^{-3} m^3)	pint (fluid), pt	0.473

	Mass		
2.20×10^{-3}	gram, g (10^{-3} kg)	pound, lb	454
3.52×10^{-2}	gram, g (10^{-3} kg)	ounce (avdp), oz	28.4
2.205	kilogram, kg	pound, lb	0.454
0.01	kilogram, kg	quintal (metric), q	100
1.10×10^{-3}	kilogram, kg	ton (2000 lb), ton	907
1.102	megagram, Mg (tonne)	ton (U.S.), ton	0.907
1.102	tonne, t	ton (U.S.), ton	0.907
	Yield and Rate		
0.893	kilogram per hectare, kg ha^{-1}	pound per acre, lb $acre^{-1}$	1.12
7.77×10^{-2}	kilogram per cubic meter, kg m^{-3}	pound per bushel, bu^{-1}	12.87
1.49×10^{-2}	kilogram per hectare, kg ha^{-1}	bushel per acre, 60 lb	67.19
1.59×10^{-2}	kilogram per hectare, kg ha^{-1}	bushel per acre, 56 lb	62.71
1.86×10^{-2}	kilogram per hectare, kg ha^{-1}	bushel per acre, 48 lb	53.75
0.107	liter per hectare, L ha^{-1}	gallon per acre	9.35
893	tonnes per hectare, t ha^{-1}	pound per acre, lb $acre^{-1}$	1.12×10^{-3}
893	megagram per hectare, Mg ha^{-1}	pound per acre, lb $acre^{-1}$	1.12×10^{-3}
0.446	megagram per hectare, Mg ha^{-1}	ton (2000 lb) per acre, ton $acre^{-1}$	2.24
2.24	meter per second, m s^{-1}	mile per hour	0.447
	Specific Surface		
10	square meter per kilogram, m^2 kg^{-1}	square centimeter per gram, cm^2 g^{-1}	0.1
1000	square meter per kilogram, m^2 kg^{-1}	square millimeter per gram, mm^2 g^{-1}	0.001
	Pressure		
9.90	megapascal, MPa (10^6 Pa)	atmosphere	0.101
10	megapascal, MPa (10^6 Pa)	bar	0.1
1.00	megagram per cubic meter, Mg m^{-3}	gram per cubic centimeter, g cm^{-3}	1.00
2.09×10^{-2}	pascal, Pa	pound per square foot, lb ft^{-2}	47.9
1.45×10^{-4}	pascal, Pa	pound per square inch, lb in^{-2}	6.90×10^3

(continued on next page)

Conversion Factors for SI and non-SI Units

To convert Column 1 into Column 2, multiply by	Column 1 SI Unit	Column 2 non-SI Unit	To convert Column 2 into Column 1, multiply by
	Temperature		
1.00 (K − 273)	Kelvin, K	Celsius, °C	1.00 (°C + 273)
(9/5 °C) + 32	Celsius, °C	Fahrenheit, °F	5/9 (°F − 32)
	Energy, Work, Quantity of Heat		
9.52×10^{-4}	joule, J	British thermal unit, Btu	1.05×10^{3}
0.239	joule, J	calorie, cal	4.19
10^{7}	joule, J	erg	10^{-7}
0.735	joule, J	foot-pound	1.36
2.387×10^{-5}	joule per square meter, J m^{-2}	calorie per square centimeter (langley)	4.19×10^{4}
10^{5}	newton, N	dyne	10^{-5}
1.43×10^{-3}	watt per square meter, W m^{-2}	calorie per square centimeter minute (irradiance), cal cm^{-2} min^{-1}	698
	Transpiration and Photosynthesis		
3.60×10^{-2}	milligram per square meter second, mg m^{-2} s^{-1}	gram per square decimeter hour, g dm^{-2} h^{-1}	27.8
5.56×10^{-3}	milligram (H_2O) per square meter second, mg m^{-2} s^{-1}	micromole (H_2O) per square centimeter second, μmol cm^{-2} s^{-1}	180
10^{-4}	milligram per square meter second, mg m^{-2} s^{-1}	milligram per square centimeter second, mg cm^{-2} s^{-1}	10^{4}
35.97	milligram per square meter second, mg m^{-2} s^{-1}	milligram per square decimeter hour, mg dm^{-2} h^{-1}	2.78×10^{-2}
	Plane Angle		
57.3	radian, rad	degrees (angle), °	1.75×10^{-2}

	Electrical Conductivity, Electricity, and Magnetism		
10	siemen per meter, $S\ m^{-1}$	millimho per centimeter, $mmho\ cm^{-1}$	0.1
10^4	tesla, T	gauss, G	10^{-4}
	Water Measurement		
9.73×10^{-3}	cubic meter, m^3	acre-inches, acre-in	102.8
9.81×10^{-3}	cubic meter per hour, $m^3\ h^{-1}$	cubic feet per second, $ft^3\ s^{-1}$	101.9
4.40	cubic meter per hour, $m^3\ h^{-1}$	U.S. gallons per minute, $gal\ min^{-1}$	0.227
8.11	hectare-meters, ha-m	acre-feet, acre-ft	0.123
97.28	hectare-meters, ha-m	acre-inches, acre-in	1.03×10^{-2}
8.1×10^{-2}	hectare-centimeters, ha-cm	acre-feet, acre-ft	12.33
	Concentrations		
1	centimole per kilogram, $cmol\ kg^{-1}$ (ion exchange capacity)	milliequivalents per 100 grams, $meq\ 100\ g^{-1}$	1
0.1	gram per kilogram, $g\ kg^{-1}$	percent, %	10
1	milligram per kilogram, $mg\ kg^{-1}$	parts per million, ppm	1
	Radioactivity		
2.7×10^{-11}	becquerel, Bq	curie, Ci	3.7×10^{10}
2.7×10^{-2}	becquerel per kilogram, $Bq\ kg^{-1}$	picocurie per gram, $pCi\ g^{-1}$	37
100	gray, Gy (absorbed dose)	rad, rd	0.01
100	sievert, Sv (equivalent dose)	rem (roentgen equivalent man)	0.01
	Plant Nutrient Conversion		
	Elemental	***Oxide***	
2.29	P	P_2O_5	0.437
1.20	K	K_2O	0.830
1.39	Ca	CaO	0.715
1.66	Mg	MgO	0.602

INTRODUCTION

Given the nature of this symposium, and the passage of a full third of a century since my brother, Bob, and I submitted our three papers on scaling, I have permitted both personal and historical material to flavor this submission quite freely, even to the extent of detailing how a personal hobby helped open the road to development. Readers preferring the filtered language of today's scientific discourse are cordially invited to thumb forward to more conventional entries.

Lesson from a Hobby. In the early 1950s, bows for hunting were often made of alloy aluminum. It was my conceit that, being a physicist, I might design an aluminum bow somewhat superior to those commercially available. To facilitate design experimentation, I wanted to use quarter-scale models, using 1/16 inch sheet stock to model the 1/4 inch sheet stock that I planned to use for my actual bow. The question was, "Can full-scale performance be deduced from small-scale models, and if so, how?" Mulling this over, I happened to invent the basic idea of similitude analysis, which was entirely unknown to me at the time.

Those aspects of physics that involve materials are normally expressed as general differential equations, and these are then integrated or solved to compute the behavior of any given macroscopic system. It dawned on me that simply by doing my scale-modeling at the level of the differential equations, I could assure that the resulting scaling of any derived systems would also hold good, even though working out the solution, mathematically, might be far too horrendous to contemplate for the given system. Such scaling of systems would be exactly as applicable as the differential equations themselves.

For my bow-and-arrow scaling, I would employ the same alloy for the full and small scale models, prestressing the bow limbs through exactly the same cycles of strain. In a super nutshell (leaving details for the well-known student), the differential equations for this problem were (i) elastic—generically, stress/strain = modulus, ($S/s = M$), and (ii) Newtonian—generically, $df = dm\ a$. Applying these conditions to arbitrary differential elements dX, dY, dZ of corresponding position and size in the two models (their respective scales being referenced to some characteristic length, L, such as the draw distance), I obtained generic results of the form $dS/dX = \rho d^2x/dt^2$ (x = a positional coordinate, X = a material coordinate). Because dS and $d(X/L)$ are the same for the two models, $dS/d(X/L)$, $L(d^2x/dt^2)$, and thus $L^2d^2(x/L)/d(t)^2$, are also the same. Therefore, $d^2(x/L)/d(t/L)^2$ is the same for both systems, meaning that time is scaled in exactly the same way as space. Consequently, mass $/L^3$, energy $/L^3$, force $/L^2$, velocity $/L^0$, and acceleration $/L^{-1}$ are the same for both the bow and its scale model.

With this result, I could now construct a miniature ballistic pendulum and use other comparisons for evaluating various designs accurately. (Incidentally, a miniature arrow flies just as fast as a full-sized arrow; when

I shot one of those toy arrows straight up, it just plain vanished into thin air—never saw it again.) After three models, the bottom line was that my final bow performed very well and actually put venison on our table. If the steel handle hadn't been so darn heavy, I might still be carrying it.

Step-by-Step Development of Soil-Moisture Scaling. A year or two after I finished that bow, Bob and I got together at the Penn State meetings where he showed me his elegant infiltration data taken at Berkeley in the wee hours, using his touchy needle-and-thread tensiometer sensor. We discussed the big differences between his sets of data for different soils. Bob pointed out that the early stages of one of the coarse soils resembled the later-stage development of one of the fine soils. We wondered whether a major portion of these differences might be simply a reflection of the average pore size of each soil. Could the differences be usefully narrowed if the complex soil moisture behavior could somehow be scale-modeled?

Back in Madison, WI, I thought about this, and played around with dimensional analysis without much luck. Eventually I tried letting gravity be zero. With this simplification, I found that infiltration distances should simply scale as the square root of time. Discussing this by phone with Bob, I learned of the Kirkham and Feng paper, which pulled together the horizontal infiltration data from many different investigators and fitted them all reasonably well to the square root of time. Showing that this result agreed with theory for such linear systems as heat flow, they remarked that because soil moisture behavior was tremendously nonlinear, there must be something interesting here that we don't yet understand. This excited me, because, of course, the result that I had deduced from dimensional analysis did not depend upon linearity. Although Arnold Klute soon published a nonlinear solution and showed that Boltzmann had also done it long before in a different context, the hook of soil physics was now set into me solidly. I never recovered from that first spurt of excitement. (Obviously, my incredible luck at having Champ Tanner dropped in my lap when he was a graduate student was also a major factor.)

Wanting to go beyond gravity-free scaling, but getting no further with dimensional analysis, one day I happened to recall my adventures with the bow and arrow, and began wondering what would happen if I tried adapting the similitude analysis idea to soils. Maybe, with similitude, I could carry scaling into the real world of gravity systems.

Fitted in between lots of other work—considerable other research, teaching, and even rebuilding part of my house at home—the work moved ahead in jerks, requiring altogether about a year. Certainly it attracted more of my interest than anything else during that time. It also led to several happy excuses for visiting Bob (who was the real soil physicist, and who was by this time located in Ithaca).

I began with the unrealistic idea of perfectly similar media, just to see where it would lead. From the beginning, I used surface tension and viscous flow (STVF) to generate the basic differential equations for similitude, accepting the consequent limitations on generality imposed by this STVF assumption.

Surface tension, with its multiple, history-dependent options of interface shape led immediately to reduce potential as $\{p\lambda/\sigma\}$ and to the time-scale invariance property of hysteresis. This was extended easily enough to the averaging of water content over large numbers of pores, yielding water content, $\theta_H\{p\lambda/\sigma\}$, thereupon defined as a scaled *hysteresis function*.

We next considered the viscous flow boundary condition for the air-water interfaces—slip, nonslip, or something in between? A lot of study, literature searching, discussion, and argument went into that phase. (Incidentally, not all of our conclusions from that aspect appeared in the final paper, having been deleted in response to the editor's request for shortening.) Basically we accepted the no-slip assumption as being simplest, and also because it evidences itself in the rising-bubble paradox, the dynamics of soap bubbles and *antibubbles*, and in other experimental observations. Adopting this boundary condition, and then averaging the net viscous flow components over many pores yielded the reduced Darcy and Richards Equations—hysteretic and in scaled form. The reduced $\{\eta K/\lambda^2\}$ was no surprise. Petroleum people always separate viscosity from their "permeability."

These averaging processes wiped away the need for "exact" similitude. If you start with one medium and then imagine shaking up a medium exactly similar to it, then repacking it to the original bulk density, it is no longer exactly similar. Yet its macroscopic properties, the θ and K hysteresis functions, are exactly the same as before the shakeup.

The final stage—using these midscale differential equations for springboarding one more step to reach the scaling of behavior for whole systems (columns, profiles)—all came tumbling out in a heap; took about an hour. It was an exciting and interesting finish with some unexpected implications. The reduced form of the gravity force required that finer-pored media must scale into larger-sized systems. Counter-intuitive.

When you mention this scale inversion effect to most nonsoil physicists, you sow the seeds of skepticism, glazing their eyeballs. (However, this glazing is also seen when you mention water under negative absolute pressure. Most of them have forgotten about Glaser's Nobel Prize for the bubble chamber, and they have never heard of Lyman Briggs.).

I remember Bob saying to me at the end of all this, "Boy, if we've slipped up on this somewhere, our necks will be out a mile." After considerable digging, we located some relevant preexisting data that supported our ideas. These went into the initial SSSA papers of 1955. Before long, Klute and Wilkinson published their spectacular confirming experiments involving both gravity and hysteresis aspects simultaneously. Then Dave Elrick and others at Wisconsin published further confirmations, and they also tested and confirmed our assumption that departures would begin to show up under conditions for which the STVF differential equations began to be inadequate for particular soils.

Insufficiency of Dimensionless Coordinates Alone. I have tried to outline how similitude analysis is based on scaling of underlying differential equations (along with their boundary conditions of course). As illustrated by our similar-medium scaling, similitude analysis is always at least as powerful as dimensional analysis and is usually more powerful. Take, for example, our micro and macro lengths, λ and L. Dimensional analysis provides no handle whatever for sorting out one of these from the other.

We can take advantage of this example to show that dimensionless coordinates, alone, need not be equivalent to scaling. Just interchanging these two lengths everywhere they occur in our list of reduced coordinates yields complete garbage. What is the meaning of $\{pL/\sigma\}$, $\{\eta K/L^2\}$, or $\{\lambda \text{div}\}$? It is as meaningful as a "purple smell' or a "bitter color." Whenever I see a paper purporting to scale some system purely in terms of cooking up dimensionless variables with no indication of their source or meaning, I wince. I have seen some of these.

Fading and Revival. The scaling theory and its initial experimental confirmations were completed by the late 1950s. Then a decade-plus passed without much being heard about scaling, though a few people were quietly using it on occasion. Eventually, the famous crew of activists at Davis began struggling with the disconcerting degree of field heterogeneity that was turning up in what seemed to me the world's most ambitious field experiments in soil physics. Claus Reichardt looked into how well such heterogeneity might be described in terms of patches of similar media with the heterogeneity residing solely in the micro-length λ. Don Nielsen claims that in the discussion following the presentation of his thesis work at one of these annual meetings, Claus found himself answering penetrating questions from two fellows in the front row. It suddenly dawned on him that these two must be Miller and Miller. It shook him up; he had assumed they were dead. Well, Don tells great stories. I won't vouch for this one, but he still sticks to it.

Abruptly, a lot of such scaling-related activity was springing up everywhere, as heterogeneity became the hot subject of the time (it still is, of course). The irony was that now I was overhearing people voicing their assumption that tests of the validity of this separate hypothesis (i.e., that heterogeneity could be described as patches of similar media) was the same thing as testing the basic validity of the scaling itself. What was being tested, of course, was whether the processes of soil genesis tend to generate patches of similar media. The testing of scaling itself that *would* prove useful (and has been almost entirely neglected) is the mapping out, over many practical soils, of the degree of limitation on the usefulness of scaling that is imposed by the original STVF assumption. Within realms of behavior that are controlled by larger pores, the STVF assumption should be valid. But just where, how, and to what degree will the limitations imposed by the very small pores of clay fractions become important?

General Perspectives. For a number of years, a large fraction of the experiments in our Soil Physics Laboratory course have been devoted to mapping out, qualitatively, the strange behavior of soils—hysteresis, infiltration, redistribution, drainage, 3D behavior, and so on. For this we have used compact, quick experiments with cheap, coarse sands. I initiated this strategy purely on the basis of scaling, since macro-size varies as the reciprocal of micro-size, and the duration of a macro-process varies as the reciprocal cube of the micro-size. (The rest of our semester largely explores practical instrumentation for finer soils and for plants, and the one-shot measurements that can be made with them in the limitations of an afternoon lab.) It works out very efficiently and I recommend it.

Go back now to our initial 1950s idea, when we were first seeking a scaling method for soils. Recall that we hoped to cut down the range of variation of soil-moisture characteristics and of flow behavior by scaling away the major effect of texture. If results are plotted in reduced coordinates, infiltration and other classes of behavior look much the same for sands as for silts, provided, of course, that everything is scaled—fluxes, total infiltration, and so on. (For practical purposes, one would also like a chance to stare at the physical coordinates; no reason that *both* can't be shown on all plotted data.)

In using this approach, generally, we must deal with all kinds of soils, hence with dissimilar soils. Plotting such data with reduced coordinates shows up the *degree* of dissimilitude, and this can be instructive. The way in which such differences are exhibited depends on the selection of a definition of micro-length λ—whether in terms of saturated conductivity, *satiated* (or rewetted) conductivity, gravity-free sorptivity, air entry value, or whatever. If we compare two similar media in terms of the ratio of their microlengths, we can choose any one of these possible definitions and the ratio of micro-lengths will always come out the same. For dissimilar media this will not be the case, and here lies a problem that generally requires further attention. The choice of definition for λ will depend on what soil property is considered most central for a given purpose. I believe that we should be thinking about such choices more than we do.

One of my relevant unfulfilled interests has been the development of better parameterized models for hysteretic soil moisture characteristics. I suggest that these parameters should, if possible, be so designed that one of them serves the role of representing microlength, somehow defined. Then, on this model, all similar media would differ only by this one parameter.

Summary. In this space I have tried to give some feeling for a small bit of history, an idea of what is—and what isn't—similitude analysis . . . how it works . . . why it is powerful. I have outlined some of the ways in which it has been useful, and a few possible ways in which it may be useful in future.

As for me, I hope it has come through how interesting this has all been. And how much fun.

EDWARD E. MILLER
University of Wisconsin
Madison, Wisconsin

1 Scaling of Freezing Phenomena in Soils[1]

R.D. Miller
Cornell University
Ithaca, New York

E.E. Miller's similitude analysis (Miller & Miller, 1956) considered capillary storage and isothermal flow of liquids in stable porous media, specifically: soils in which grain sizes are large enough that surface adsorption effects can reasonably be neglected—except as they determine wetting angles—but fine enough to allow neglect of momentum terms in applications of the Navier-Stokes equation to microscopic flow patterns in the interstices. Such soils have been called *STVF soils* to signify that pore liquids obey the laws of surface tension and viscous flow (Miller, 1980a).

The analysis produced a set of dimensionless, fully reduced versions of physical variables pertaining to such processes as infiltration and drainage. Their definitions implied the rules for construction and operation of three distinct classes of scale models of STVF prototype systems: *population analogues* (coarser or finer than the prototype), *substance analogues* (different pore liquids), and *bodyforce analogues* (different bodyforce fields).

Many soils that qualify as STVF media are subject to processes known as *frost heaving* and *freezing-induced redistribution*. Both are important thermomechanical processes, but both are awkward to study in the field. Field problems are difficult to simulate in the laboratory. Computer simulations can be instructive but difficult to validate and it is difficult to obtain data needed to properly characterize even a simple unstratified soil.

Can Miller's (1980a) reduced variables provide a basis for interpolations and extrapolations of available field observations or laboratory tests? Is there a future for scale modeling of important problems, using rules like those implied by Miller's (1980a) definitions of reduced variables for capillary storage and isothermal flow? This chapter will suggest what may be possible and what is not.

---

[1]Contribution from the Department of Agronomy, Cornell University, Ithaca, NY.

LIQUISCOPIC GRAINS

One way to construct a rational explanation of frost heaving is to imagine that any tiny parcel of liquid water that is very close to the surface of a soil grain will experience an intense body force, a force that diminishes rapidly with distance, becoming imperceptible at distances of, say, a few tens of molecular diameters from the grain surface. At the same time, one imagines that neither ice nor air experiences comparable body forces and, consequently—being held at a distance from a grain surface by an intervening adsorbed film of mobile liquid—ice and air may not experience any adsorption force at all! The term *hygroscopic* brings to mind a surface that has an affinity for molecular H_2O, not necessarily in a liquid state. The term *liquiscopic* has been coined for an idealized affinity that is specific for the *liquid* phase, whether aqueous or not. Having imagined this situation, and its deduced consequences, one can then explore mechanisms that would either explain such a bodyforce (such as a model based on the diffuse electrical double layer), or any other mechanism that would appear to produce equivalent behavior. But that will be left to another occasion.

A liquiscopic STVF soil will be called an *LSTVF soil.* In such a soil, space close to a grain surface, where a tiny parcel of liquid would experience a perceptible adsorption force, will be called *adsorption space.* The remaining interstitial space will be called *capillary space.* For a soil to qualify as an LSTVF soil, the volume of adsorption space must be small compared to capillary space if pores are to be assumed to be stable.

MECHANISMS

It is often remarked that explanations of freezing phenomena in soils are "controversial." As one whose views have contributed to controversies, the author sees freezing-induced redistribution and frost heaving as logical consequences of the laws of surface tension and viscous flow in liquiscopic media with phase transformations obeying the laws of thermodynamics as embodied in a generalized form of the Clapeyron equation. Some of those consequences can be dealt with using concepts of soil mechanics as represented by Terzaghi's effective stress equation and its derivatives. Other parties to the controversies favor a "hydrodynamic" model, explicitly or implicitly bypassing consideration of the physics of freezing at the microscopic level, especially implications reflecting application of the laws of surface tension to ice-water and ice-air interfaces.

METHOD

A formal similitude analysis of freezing phenomena in LSTVF soils would be largely redundant. Miller's (1980a) results for the laws of surface tension and viscous flow can simply be accepted and recast with modifications appropriate for multiphase systems in which phase transitions obey the laws of thermodynamics. From then on, with specific conceptions of behavior compatible with the laws of surface tension, viscous flow, diffusion of sensible heat, and surface adsorption phenomena in mind, it is a matter of arranging the relevant physical equations in such a sequence that each equation contains, in addition to variables for which reduced versions have already been defined, one or more additional variables whose definitions must be compatible with the antecedent definitions. Compatible definitions can then be formally declared before going on to the next equation.

Equations in this chapter are applicable to freezing-induced redistribution as a capillary sink phenomenon (Dirksen & Miller, 1966; Miller, 1973; Bresler & Miller, 1975; Colbeck, 1982), and to frost heaving as a thermally induced regelation phenomenon (Römkens & Miller, 1973) as embodied in the rigid ice model of heaving (Miller, 1978, 1980b; Snyder & Miller, 1985; O'Neill & Miller, 1985; Black & Miller, 1985).

Liquiscopicity of a surface plays only a passive role in freezing-induced redistribution, preventing ice-water and water-air interfaces from intersecting grain surfaces. If no heaving is taking place, redistribution takes place in capillary space where reduced variables will be compatible with those taken from Miller's (1980a) original similitude analysis.

Liquiscopicity of grain surfaces plays an active role in thermally induced regelation, an adsorption-space phenomenon that is linked to a chain of simultaneous events that are taking place in capillary space. Liquiscopicity, therefore, plays a vital but not necessarily dominant role in the mechanism of frost heaving.

SYMBOLS

This section lists and defines all symbols that will turn up in the equations listed in the succeeding section. SI units are noted for those symbols represented by lower case Greek and Roman letters. Dimensionless ratios are represented by upper case letters.

Subscripts

G, W, I, A	*Grains*, *water*, *ice*, *air*, or their analogues.
H	Sensible *heat*.
R	"Regelation."

Symbols, Units, and Definitions

η	$\mathrm{N\ m^{-2}\ s}$	Viscosity of water or its analogue.
f_W	$\mathrm{N\ m^{-3}}$	Body force per unit volume of water, or its analogue, due to gravity or centrifugation.

$\gamma_{IW}, \gamma_{WA}, \gamma_{IA}$	$J\ m^{-2}$	Specific surface free energies ("surface tensions") of, respectively, ice-water, water-air and ice-air interfaces or their analogues.
a		Angle of convergence of ice-water and water-air interfaces at a transition to ice-air interface. Small, perhaps zero, for H_2O.
θ_o	k	Standard melting point of bulk ice, or its analogue, when exposed to air at standard atmospheric pressure.
θ	k	Elevation of melting point of ice, or its analogue, with respect to its melting point under "standard conditions." For H_2O, θ will be temperature in degrees Celsius, by definition. For H_2O, θ will generally be negative.
h_{IW}	$J\ m^{-3}$	Latent heat of fusion, per unit volume of melt.
Y_{IW}		Density ratio, ice/water, or their analogues.
p_I, p_W, p_A	$N\ m^{-2}$	Absolute pressures of, respectively, ice, water and air, or their analogues.
p_o	$N\ m^{-2}$	Standard atmospheric pressure.
u_I, u_W, u_A	$N\ m^{-2}$	Standard gauge pressures of, respectively, ice, water, and air, or their analogues, in capillary space; $u_x \equiv p_x - p_o$.
$\phi_{IW} \equiv (u_I - u_W)$	$N\ m^{-2}$	Phi-variable. In capillary space, $\phi_{IW} \equiv (u_I - u_W)$
$\phi_{WA} \equiv (u_W - u_A)$	$N\ m^{-2}$	Phi-variable. In capillary space, $\phi_{WA} \equiv (u_W - u_A)$ ϕ_{WA} is equivalent to "matric potential."

$\phi_{IA} \equiv (u_I - u_A) \equiv (\phi_{IW} + \phi_{WA})$	N m^{-2}	Phi-variable. In capillary space. $\phi_{IA} \equiv (u_I - u_A) = (\phi_{IW} + \phi_{WA})$.
r_{IW}, r_{WA}, r_{IA}	m	Mean radii of curvature of, respectively, ice-water, water-air, and ice-air interfaces, or their analogues.
σ, σ_e, σ_n	N m^{-2}	Terzaghi's "total stress" (envelope pressure), "effective stress" and "neutral stress" at the level of incipient lens initiation (includes weight of overburden).
σ_e^*, σ_n^*	N m^{-2}	Effective stress and neutral stress at a site of incipient lens initiation (as in Snyder's version of "flawed solid" theory of tensile failure for moist soils).
v_R	m s^{-1}	Microscopic velocity of ice lattice (or its analogue) relative to grains in the "frozen fringe" below a growing ice lens. Numerically equal to rate of heave in LSTVF soil.
t	s	Time.
v_W, v_I	m^3 m^{-2} s^{-1}	Volumetric fluxes of, respectively, water and ice, or their analogues.
v_H	W m^{-2}	Macroscopic flux of sensible heat (convective transport by moving water neglected).
λ	m	A microscopic length that characterizes the coarseness of a grain population.
ξ	m	A macroscopic length used to characterize the size of a system in the sense of corresponding distances in a model and its prototype.
div	m^{-1}	Divergence operator.
grad	m^{-1}	Gradient operator.

Characteristics of a Given LSTVF Soil, Mainly Hysteretic Functions

$W(\phi_{IW}, \phi_{WA})$		Volume fraction of liquid water or its analogue.
$I(\phi_{IW}, \phi_{WA})$		Volume fraction of ice or its analogue.
$A(\phi_{IW}, \phi_{WA})$		Volume fraction of air or its analogue.
$X(\phi_{IW}, \phi_{WA})$		Bishop's factor for partitioning neutral stress between two pore phases at unequal pressures.
$X^*(\phi_{IW}, \phi_{WA})$		Snyder's factor for partitioning neutral stress between two pore phases at unequal pressures at the site of tensile failure in an unsaturated soil or at the site of lens initiation in a freezing soil.
$k_W(\phi_{IW}, \phi_{WA})$	$m^4\ N^{-1}\ s^{-1}$	Capillary conductivity (Darcy coefficient), liquid flow only.
$k_H(\phi_{IW}, \phi_{WA})$	$W\ m^{-1}\ k^{-1}$	Thermal conductivity (Fourier coefficient) thermal diffusion only.
$c_H(\phi_{IW}, \phi_{WA})$	$J\ m^{-3}\ k^{-1}$	Volumetric heat capacity of the medium, including pore phases, excluding heat adsorbed or released as a result of phase changes associated with temperature changes.
G		Volume fraction of grains; invariant in a given LSTVF soil.

EQUATIONS AND REDUCED VARIABLES

This section lists equations that enter into physical models of redistribution and heaving using the symbols defined in the previous section. If an equation permits a previously undefined reduced variable to be defined, the appropriate definition is then stated, whereupon the initial equation is restated in reduced form using the corresponding upper case Greek and Roman letters for the reduced versions of the physical variables.

a. $\gamma_{IW} + \gamma_{WA} = \gamma_{IA} \cos a$ Antonov's Rule

Define:

$$\Gamma_{IW} \equiv [(1/\gamma_{IA})\gamma_{IW}];\ \Gamma_{WA} \equiv [(1/\gamma_{IA})\gamma_{WA}]$$

A. $\Gamma_{IW} + \Gamma_{WA} = \cos a$

b. $p_1 - p_W = 2\ \gamma_{IW}/r_{IW}$ Laplace Equations

$$p_W - p_A = 2\ \gamma_{WA}/r_{WA}$$

$$p_1 - p_A = 2\ \gamma_{IA}/r_{IA}$$

Define:

$$R_{IW} \equiv [(1/\lambda)r_{IW}];\ R_{WA} \equiv [(1/\lambda)r_{WA}];\ R_{IA} \equiv [(1/\lambda)r_{IA}]$$

$$P_I \equiv [(\lambda/\gamma_{IA})p_I];\ P_W \equiv [(\lambda/\gamma_{IA})p_W];\ P_A \equiv [(\lambda/\gamma_{IA})p_A]$$

B. $P_I - P_W = 2\Gamma_{IW}/R_{IW}$

$$P_W - P_A = 2\Gamma_{WA}/R_{WA}$$

$$P_I - P_A = 2/R_{IA} = 2(\Gamma_{IW}/R_{IW} + \Gamma_{WA}/R_{WA})$$

c. $u_w - u_I/Y_{IW} = (h_{IW}/\theta_o)\theta$ Clapeyron Equation

Define:

$$H_{IW} \equiv [(\lambda/\gamma_{IA})\ h_{IW}];\ \Theta \equiv [(1/\theta_o)\theta]$$

C. $U_W - U_i/Y_{IW} = H_{IW}\ \Theta$

d. $G + W(\phi_{IW}, \phi_{WA}) + I(\phi_{IW}, \phi_{WA})$ Sum of Volume Fractions

$$+ A(\phi_{IW}, \phi_{WA}) = 1.0$$

Define:

$$\Phi_{IW} \equiv [(\lambda/\gamma_{IA})\phi_{IW}];\ \Phi_{WA} \equiv [(\lambda/\gamma_{IA})\phi_{WA}];\ \Phi_{IA} \equiv [(\lambda/\gamma_{IA})\phi_{IA}]$$

D. $G + W(\Phi_{IW}, \Phi_{WA}) + I(\Phi_{IW}, \Phi_{WA}) + A(\Phi_{IW}, \Phi_{WA}) = 1.0$

e. $\sigma = \sigma_e^* + \sigma_n^*$ Snyder-Terzaghi Equation

Define:

$$\Sigma \equiv [(\lambda/\gamma_{IA})\sigma];\ \Sigma_e^* \equiv [(\lambda/\gamma_{IA})\sigma_e^*];\ \Sigma_n^* \equiv [(\lambda/\gamma_{IA})\sigma_n^*]$$

E. $\Sigma = \Sigma_e^* + \Sigma_n^*$

f. $\sigma_n^* = X^*(\theta_{IW}, \theta_{WA})u_W + [1 - X^*(\theta_{IW}, \theta_{WA})]u_I$ Snyder-Bishop Equation

F. $\Sigma_n^* = X^*(\Phi_{IW}, \Phi_{WA})U_W + [1 - X^*(\Phi_{IW}, \Phi_{WA})]U_I$

g. $X^*(\phi_m, 0) = \frac{1}{2}\left[S(\phi_m, 0) - (0.3/\phi_m)\sum_{j=0}^{m}\phi_j\, dS\right]$ Snyder-Aitchison Equation

G. $X^*(\Phi_m, 0) = \frac{1}{2}\left[S(\Phi_m, 0) - (0.3/\Phi_m)\sum_{j=0}^{m}\Phi_j\, dS\right]$

h. $\sigma_e^* = 0$ Condition for Initiation of an Ice Lens

H. $\Sigma_e^* = 0$

i. $v_w = k_w(\phi_{IW}, \phi_{WA})\,(f_W - \text{grad}\, u_W)$ Darcy's Law

Define:

$$\text{GRAD} \equiv (\xi\ \text{grad})$$

$$F_W \equiv [(\xi\lambda/\gamma_{IA})f_W]$$

$$K_W \equiv [(\eta/\lambda^2)k_w]$$

$$V_W \equiv [(\xi\eta/\lambda\gamma_{IA})\,v_W]$$

$$V_W = K_W(\Phi_{IW}, \Phi_{WA})(F_W - \text{GRAD } U_W)$$

j. $\text{div}(v_W + Y_{IW}v_I) = -\frac{\delta}{\delta t}[W(\phi_{IW}, \phi_{WA}) + Y_{IW} I(\phi_{IW}, \phi_{WA})]$ — Conserve Pore Substance

Define:

$$\text{DIV} \equiv (\xi \text{ div})$$

$$V_I \equiv [(\xi\eta/\lambda\gamma_{IA})v_I]$$

$$T \equiv [(\lambda\gamma_{IA}/\xi^2\eta)t]$$

J. $\text{DIV}(V_W + Y_{IW}V_I) = -\frac{\partial}{\partial T}[W(\Phi_{IW}, \Phi_{WA}) + Y_{IW}I(\Phi_{IW}, \Phi_{WA})]$

k. $v_I = I(\phi_{IW}, \phi_{WA})\, v_R$ — Flux of (Rigid) Ice by Regelation

Define:

$$V_R \equiv [(\xi\eta/\lambda\gamma_{IA})v_R]$$

K. $V_I = I(\Phi_{IW}, \Phi_{WA})V_R$

l. $\text{div}(v_H + h_{IW}v_W) = -\frac{\partial}{\partial t}[c_H(\phi_{IW}, \phi_{WA})\theta + h_{IW} W(\phi_{IW}, \phi_{WA})]$ — Conserve Thermal Energy

Define:

$$V_H \equiv [(\xi\eta/\gamma_{IA}^2)v_H]$$

$$C_H(\Phi_{IW}, \Phi_{WA}) \equiv [(\lambda\theta_o/\gamma_{IA})c_H(\Phi_{IW}, \Phi_{WA})]$$

L. $\text{DIV}(V_H + H_{IW} V_W) = -\frac{\delta}{\delta T}[C_H(\Phi_{IW}, \Phi_{WA})\Theta + H_{IW} W(\Phi_{IW}, \Phi_{WA})]$

m. $v_H = -k_H(\phi_{IW}, \phi_{WA})\text{grad } \theta$ — Fourier's Law

Define:

$$K_H(\Phi_{IW}, \Phi_{WA}) \equiv [(\xi\theta_o/\gamma_{IA}^2)k_H(\phi_{IW}, \phi_{WA})]$$

M. $V_H = -K_H(\Phi_{IW}, \Phi_{WA})\text{GRAD }\Theta$

CONCLUSIONS

The rules for construction and operation of scale-models, implicit in the definitions of reduced variables presented above, preclude consideration of substance analogues for modeling either freezing-induced redistribution or frost heaving in an aqueous prototype. One cannot expect to find any real substance for which reduced variables relating to densities, fractional surface tensions, heats of fusion, heat capacities, and thermal conductivities all match those of an aqueous prototype, even approximately. This conclusion does not preclude the possibility of analogous processes involving other pore substances, but they cannot serve as scale-models of freezing phenomena in aqueous prototypes.

The definitions suggest that if heaving is suppressed, a population analogue is a promising candidate for scale-modeling of freezing-induced redistribution. On the other hand, expectations of imperfect similitude in adsorption space impair expectations for acceptable performance of a population analogue for scale-modeling of frost heaving in a prototype population.

For scale-modeling of freezing-induced redistribution and heaving, a bodyforce analogue suffers none of the obvious disabilities of substance and population analogues. Although the question has not been examined in this chapter, this conclusion should apply to compressible soils in which the volume of adsorption space is not limited to a small fraction of the total space as in the case of LSTVF soils.

The bodyforce analogue is by far the most promising approach to the scale-modeling of problems that involve prolonged freezing of an aqueous prototype, including cumulative effects of what may become substantial weights of frozen overburden. An example of such a problem is self-induced heaving of buried large-diameter natural gas pipelines traversing windows in permafrost. Necessarily operated at subfreezing temperatures to avoid thawing of permafrost on either side of such windows, such a pipeline may self-destruct as a result of self-induced heaving that may continue throughout the lifetime of the system. In principle, a geotechnical centrifuge capable of spinning a cubic meter of soil at 100 Gs could, in one day, using a model built at a scale ratio of 1:100, similute—at that same scale—25 yr of freezing and heaving of a 100-m section of such a pipeline buried in that same soil and operated at the same temperatures.

The idea of using a geotechnical centrifuge to model geotechnical problems in frozen or freezing ground can be found in the Soviet literature

(Pokrovsky & Fyodorov, 1969). It was not explicitly suggested, however, that it might be applicable to scale-modeling of frost heaving or freezing induced redistribution.

REFERENCES

Black, P.B., and R.D. Miller. 1985. A continuum approach to modeling of frost heaving. p. 36–45. *In* D.M. Anderson and P.J. Williams (ed.) Freezing and thawing of soil-water systems. Am. Soc. of Civil Eng., New York.

Bresler, E., and R.D. Miller. 1975. Estimation of pore blockage induced by freezing of unsaturated soil. p. 161–175. *In* J.N. Luthin (chair) Proc. Conf. on Soil-Water Problems in Cold Regions. Calgary, Alberta. 6–7 May. Spec. Task Force Am. Geophys. Union. James N. Luthin, Univ. of California, Davis.

Colbeck, S.C. 1982. Configuration of ice in porous media. Soil Sci. 133:116–123.

Dirksen, C., and R.D. Miller. 1966. Closed-system freezing of unsaturated soil. Soil Sci. Soc. Am. Proc. 30:168–173.

Miller, E.E. 1980a. Similitude and scaling of soil-water phenomena. p. 300–318. *In* D. Hillel (ed.) Applications of soil physics. Academic Press, New York.

Miller, E.E., and R.D. Miller. 1956. Physical theory for capillary flow phenomena. J. Appl. Phys. 4:324–332.

Miller, R.D. 1973. Soil freezing in relation to pore water pressure and temperature. p. 344–352. *In* Permafrost, North Am. Contrib., 2nd Int. Conf., Yakutsk, USSR. 13–28 July. Nat. Acad. Sci. Washington, DC.

Miller, R.D. 1978. Frost heaving in non-colloidal soils. p. 707–713. *In* Proc. 3rd. Int. Conf. on Permafrost. Vol 1. Edmonton, Alberta 10–13 July. Nat. Res. Council of Canada, Ottawa.

Miller, R.D. 1980b. Freezing phenomena in soils. p. 300–318. *In* D. Hillel (ed.) Applications of soil physics. Academic Press, New York.

O'Neill, K., and R.D. Miller. 1985. Exploration of a rigid ice model of frost heaving. Water Resour. Res. 21:281–296.

Pokrovsky, G.I., and I.S. Fyodorov. 1969. Centrifugal model testing in the construction industry. Draft translation prepared by Building Res. Estab., UK.

Römkens, M.J.M., and R.D. Miller. 1973. Migration of mineral particles in ice with a temperature gradient. J. Colloid Interface Sci. 42:103–111.

Snyder, V.A., and R.D. Miller. 1985. Tensile strength of unsaturated soils. Soil Sci. Soc. Am. J. 49:58–65.

2 Miller Similitude and Generalized Scaling Analysis

Garrison Sposito
University of California
Berkeley, California

William A. Jury
University of California
Riverside, California

Miller similitude is a physically based algorithm for defining scale-invariant relationships concerning the properties of water in homogeneous porous media (Miller & Miller, 1956; Miller, 1980). The fundamental concept underlying this algorithm is that of a characteristic length scale that reflects the sizes of solid particles and the dimensions of pores in a particular geometric arrangement. Similitude then results from the use of this length scale as a factor to render transport coefficients and potentials for water in porous media in a scaled form. If two porous media exhibit the same scaled relationships among their transport coefficients and water potentials, they are said to be *Miller-similar*, and if they share identical values of the scaled properties, they are in *Miller-similar states*.

Table 2–1 summarizes the basic features of Miller similitude as applied to the volumetric water content, matric potential, and hydraulic conductivity. In each example, the asterisk subscript denotes a scaled form of a medium-water property. Similar media are those that show the same functional dependence of one scaled property on another [e.g., the same scaled moisture characteristic, $\theta_*(\Psi_*)$].

Miller (1980) has discussed the application of Miller similitude to the behavior of water in soil, and Sposito and Jury (1985) have compared Miller similitude with the strictly mathematical technique of inspectional analysis (Tillotson & Nielsen, 1984), which is used to cast physical equations into dimensionless form. Perhaps the most ambitious application of the relationships in Table 2–1 is their use to characterize water in field soils (Warrick & Nielsen, 1980; see chapter 4 in this book). In this case, one

 Scaling in Soil Physics: Principles and Applications, SSSA Special Publication no. 25.

Table 2–1. Scaling relations in classical Miller similitude.

Soil water property	Symbol and dimensions	Scaling relation†
Volumetric water content	$\theta[L^3L^{-3}]$	$\theta_* = \theta$
Matric potential	$\psi[L]$	$\psi_* = \lambda\psi$
Hydraulic conductivity	$K[LT^{-1}]$	$K_* = K/\lambda^2$

†Asterisk denotes the scaled property that is the same for all Miller-similar porous media; λ denotes a characteristic length.

imagines that a heterogeneous field soil is a union of approximately homogeneous domains, each of which can be represented by a single characteristic length scale. Heterogeneity then is reflected in the spatial variability of the length scale, and, if the homogeneous domains are Miller-similar, the functional relationships among soil-water properties will be uniform. For example, if $K(\Psi)$ represents the dependence of the hydraulic conductivity on the matric potential in the neighborhood of some point in a soil and $K'(\Psi')$ does the same at another point, then the physical scaling relationship

$$\frac{K'(\lambda'\Psi')}{\lambda'^2} = \frac{K(\lambda\Psi)}{\lambda^2} \qquad [1]$$

where λ is a characteristic length, can be used to relate the two local hydraulic conductivities, if the functional dependence of each on Ψ has the same mathematical form. If this kind of scaling relationship holds, the theory of water movement in field soils is greatly simplified, because the explicit dependence of soil-water properties on spatial position has been removed.

The purpose of this chapter is to explore further the possibility that scaling analysis can be applied effectively to characterize the dynamics of water in field soils. The success of scaling techniques related to inspectional analysis (Sposito & Jury, 1985) and the difficulties that have been encountered in applying Miller similitude to field soils (Jury et al., 1987) suggest that a generalization of the latter in the direction of the former might prove fruitful. A physically based, generalized scaling analysis will be developed in this chapter and applied to the equation of motion for water in soil (the Richards equation) as well as to the rigorous statistical characterization of soil-water scale factors in heterogeneous field soils.

GENERALIZED SCALING ANALYSIS

Classical Miller similitude is founded on a physical relationship between effective pore size and matric potential in a porous medium (Miller & Miller, 1956). If $\Psi[L]$ is the matric potential to which corresponds the effective pore radius $r[L]$, then

$$\Psi(r) \equiv 2\sigma/\rho g r \qquad [2]$$

where $\sigma[MT^{-2}]$ is the surface tension at the air-water interface, $\rho[ML^{-3}]$ is the mass density of liquid water, and $g[LT^{-2}]$ is the gravitational acceleration. Equation [2] defines a pore radius corresponding to each measured value of $\Psi > 0$. Two porous media are said to be Miller-similar if a scale factor $\lambda[L]$ can be found for each such that identical values of the dimensionless ratio r/λ imply geometric similitude in the microscopic arrangement of pores and solid particles (Miller & Miller, 1956; Miller, 1980). Similar media of this type should exhibit microscopic structures that look identical, except for scale, in the same sense as do similar triangles.

It follows from Eq. [2] and the definition of *similar* that similar porous media should be in similar states (Miller & Miller, 1956) whenever they exhibit the same values of the product $\lambda\Psi\rho/\sigma$, with each λ involved being a characteristic length for a given medium. Miller and Miller (1956) postulated that porous media in similar states should have the same volumetric water content θ and, therefore, that the moisture characteristic for similar media should have a common mathematical form when expressed as $\theta(\lambda\Psi\rho/\sigma)$. Experimental tests of this postulate, although limited in number, indicate that it describes well-sorted sands reasonably well, but applies poorly to soils containing a broad range of particle sizes (Miller, 1980; Tillotson & Nielsen, 1984). In field soils, wherein the porosity varies from place to place in an unpredictable manner, the postulate is a priori invalid (Jury et al., 1987).

The concept of similar state based on Eq. [2] actually does not require the postulate of invariance of the volumetric water content under the scaling of Ψ. A valid alternative, for example, is to postulate the invariance of the pore-size distribution instead. The conceptual difference involved has to do with the physical interpretation of the scale factor λ. If the water content (or, equivalently, the porosity) is an invariant of scaling, λ necessarily reflects the geometric arranagement of both the pore space and the solid particles, because θ is the ratio of a pore space volume to a porous medium volume. On the other hand, if the pore-size distribution is the invariant quantity, then λ is associated only with the pore space, as implied already in Eq. [2], and the possibility exists that the solid particle arrangement could have a different scale factor.

The pore-size distribution is measured conventionally (Danielson & Sutherland, 1986) by determining the relative water saturation, $\theta/\theta_s \equiv S$, as a function of Ψ, where θ_s is the volumetric water content at saturation. A graph of S vs. r as calculated with Eq. [2] then estimates the cumulative pore-size distribution function for a porous medium. In terms of this operation, the postulate of invariance of the pore-size distribution under scaling implies that similar media will show the same mathematical form for $S(\lambda\Psi\rho/\sigma)$. This weaker concept of similitude is compatible with a broad range of particle size and nonuniform porosity. Field studies of macroscopic similitude (Warrick & Nielsen, 1980; Jury et al., 1987) have shown that S is a useful variable in terms of which to express scaled soil water properties. Its postulated invariance under scaling of the matric potential and the theory of similitude predicated on this invariance will be termed *generalized scaling analysis* to distinguish it from classical Miller scaling analysis. Clearly, the latter implies the former, but not the converse.

SCALING INVARIANCE OF THE RICHARDS EQUATION

The physics of infiltration in porous media can be described in its simplest aspects by a nonlinear partial differential equation known as the Richards equation (see, e.g., Hillel, 1980; Sposito, 1986). In one spatial dimension, this evolution equation takes the form

$$\frac{\partial \theta}{\partial t} = \frac{\partial}{\partial z}\left[D(\theta)\frac{\partial \theta}{\partial z}\right] - \frac{\partial K}{\partial \theta}\frac{\partial \theta}{\partial z} \qquad (t > 0) \qquad [3a]$$

where $K[LT^{-1}]$ is the hydraulic conductivity,

$$D \equiv K\frac{\partial \Psi}{\partial \theta} \qquad [4]$$

is the water diffusivity (with dimensions L^2T^{-1}), and the coordinate z is restricted to nonnegative values. The invariance of Eq. [3] under scaling transformations has been investigated by Sposito (1990) using the theory of Lie groups. This approach, whose results will be only summarized here, provides a systematic framework within which to classify differing perspectives on the scaling of soil water properties (Sposito & Jury, 1985).

The scaling transformations of Eq. [3] are perhaps exposed most clearly if "natural" dependent and independent variables are chosen so as to cast the Richards equation into dimensionless form. The natural dependent variable is the relative water saturation, S. A natural length variable can be defined by a chosen value of Ψ (Simmons et al., 1979; Youngs & Price, 1981; Sposito & Jury, 1985), and a corresponding time variable is defined by the ratio of the chosen Ψ-value to the hydraulic conductivity at water saturation, $K_s \equiv K(\theta_s)$. Thus, Eq. [3a] can be put into dimensionless form with the set of transformations:

$$S \equiv \theta/\theta_s, \quad T \equiv K_s t/\theta_s \Psi_o, \quad X \equiv z/\Psi_o, \qquad [5]$$

such that the Richards equation becomes

$$\frac{\partial S}{\partial T} = \frac{\partial}{\partial x}\left[d(S)\frac{\partial S}{\partial x}\right] - \frac{\partial k}{\partial S}\frac{\partial S}{\partial x} \qquad (T > 0) \qquad [3b]$$

where

$$d(S) \equiv D(S)/\Psi_o K_s \qquad k(S) \equiv K(S)/K_s \qquad [6]$$

are dimensionless transport coefficients and $S < 1$. Equation [3b] can serve as the point of departure for a Lie group analysis of scale invariance.

If a porous medium to which Eq. [3b] applies, under suitable initial and boundary conditions, shows generalized scaling invariance for soil water properties, it is said to be *Warrick-similar* (Sposito & Jury, 1985). The scale transformations in generalized Warrick similitude are (Sposito, 1990):

$$S' = \mu S + \delta \qquad T' = a\omega^2 T \qquad X' = aX \tag{7}$$

where μ, a, and ω are positive scale factors and δ is a translation parameter. It is straightforward to show (Sposito & Jury, 1985; Sposito, 1990) that a and μ/ω^2 are scaling parameters for the matric potential and hydraulic conductivity, respectively:

$$h' = ah \qquad k' = \mu k/\omega^2 \tag{8}$$

where

$$h(S) \equiv \Psi(S)/\Psi_o \tag{9}$$

is a dimensionless matric potential. Lie group analysis applied to Eq. [3b] and [7] leads to the following general conclusions (Sposito, 1990):

1. The transport coefficients can depend on the relative water saturation in only two ways ($S < 1$)

$$d(S) = (pS + q)^m \tag{10a}$$

$$k(S) = [c_1/p(1 + mB)][(pS + q)^{1+mB}] + c_0 \tag{10b}$$

$$d(S) = A\exp(bS) \tag{11a}$$

$$k(S) = (c_2/Bb)\exp(BbS) + c_0 \tag{11b}$$

where p, q, m, c_0, c_1, c_2, A, and b are arbitrary constants and the positive parameter B is defined by the equation

$$\frac{d(S)}{\partial d/\partial S} = B\frac{\partial k/\partial S}{\partial^2 k/\partial S^2} \qquad (B \neq 0.5 \text{ or } 1) \tag{12}$$

Equations [10] and [11] show that the transport coefficients can have only a power-law or an exponential dependence on S if the Richards equation is to be invariant under scale transformations.

2. The scale factors are interrelated by the equations

$$a = \omega^{2(1-1/B)} \tag{13a}$$

$$\mu = (a/\omega^2)^{1/m} \text{ (power-law)} \qquad \mu = 1 \text{ (exponential)} \tag{13b}$$

Thus, $a = \omega$ only if $B = 2$: equality of the scale factors for Ψ and K occurs in a single case of Eq. [12]. Moreover, it follows from Eq. [8] and [13] that

$$\frac{k}{k'} = \left(\frac{h}{h'}\right)^{-\eta} \qquad [14]$$

is also a general condition for scale-invariance of the Richards equation, where

$$\eta \equiv \frac{B + {}^{1}/m}{B - 1} \quad \text{(power law)} \qquad [15a]$$

$$\eta \equiv B/(B - 1) \text{ (exponential)} \qquad [15b]$$

and $B \neq 0.5$ or 1 (Sposito, 1990).

The physical significance of the result (1.) above is that the Richards equation will be invariant under scaling transformations only if the water diffusivity and hydraulic conductivity have the mathematical forms specified in Eq. [10] and [11]. This condition adds to the requirement for similar porous media the need to establish either a power-law or an exponential dependence of the transport coefficients on the water content. The same can be shown to be true for the matric potential (Sposito, 1990). If the transformation in Eq. [7] is interpreted as a relation between two locally homogeneous domains in a field soil [cf. Sposito & Jury (1986)], then (1.) above indicates that each domain must show soil-water properties with the simple dependence on water content in Eq. [10] or [11], if the Richards equation is to apply to both via a scaling relationship. Once this condition is verified experimentally, it is necessary to solve the Richards equation only for one such local domain; the solution then can be scaled to apply it to other domains in the field soil (Warrick & Nielsen, 1980; Warrick et al., 1985).

The result (2.) above shows that the scale factors for the matric potential and hydraulic conductivity in general are not equal, i.e., $a \neq \omega$ unless $B = 2$ in Eq. [12]. This condition also may be checked experimentally for a local homogeneous region in a field soil. Classical Miller similitude requires $a = \omega/\mu^{1/2}$, as implied in Table 2–1. Sposito and Jury (1985) and Jury et al. (1987) have reviewed the evidence for equality, pointing out that a and ω usually are observed to be correlated, instead of equal, in field studies of soil water properties. Generalized scaling analysis shows that neither relationship is a requirement for the scaling invariance of the Richards equation.

Irrespective of whether $a = \omega/\mu^{1/2}$, scaling invariance does require that pairs of hydraulic conductivity/matric potential values related via the transformations in Eq. [7] satisfy the inverse power-law expression in Eq. [14], regardless of the S-dependence in Eq. [10] and [11]. Note that $B =$

2 leads to $\eta = 2$ in Eq. [15b] and $\eta = 2 + m^{-1}$ in Eq. [15a]. All of these constraints on the scaling of the Richards equation can, in principle, be tested experimentally.

EXPERIMENTAL TESTS ON FIELD SOILS

Jury et al. (1987) have conducted a thorough examination of the scaling relationships for $h(S)$ and $k(S)$ using large databases for two field soils developed by Nielsen et al. (1973) and Russo and Bresler (1981). The study of Russo and Bresler (1981) involved measurements of θ, θ_s, Ψ, Ψ_0 ("air entry value"), K, and K_s in experimental plots distributed over a 0.8-ha field containing a heterogeneous Rhodoxeralf (Hamra Red Mediterranean soil). A total of 120 data sets was developed for the 0- to 0.9-m depth in the soil. The data were fit to the power-law relationships:

$$h(S) = S^{-m(B-1)} \qquad k(S) = S^{1+mB} \qquad [16]$$

consistently with Eq. [10]. [Russo & Bresler (1981) used notation differing from that in Eq. [16].] The field-wide mean values of m and B were 1.676 and 1.404, respectively. The study of Nielsen et al. (1973) involved measurements of soil water properties in irrigated plots scattered over a 150-ha field containing a heterogeneous Entisol [Panoche series; fine-loamy, mixed (calcareous), thermic Typic Torriorthents]. A total of 120 data sets was developed for the 0.30- to 1.83-m depths in the soil. The data for ψ and K as a function of θ have been fit to exponential relationships by Nielsen et al. (1973) and Simmons et al. (1979). Jury et al. (1987) used the data to calculate $h(S)$ and $k(S)$ [with $\psi_0 \equiv \psi$ (0.98)]. Rounded field-wide mean values and coefficients of variation (CV) for $h(S)$ and $k(S)$, computed for $0.75 \leq S \leq 0.975$ by Jury et al. (1987), are summarized in Table 2–2 for both soils.

Scale factors for $h(S)$ and $k(S)$ were defined by Jury et al. (1987) to

Table 2–2. Statistical analysis† of $h(S)$ and $k(S)$ in two field soils.

Relative water saturation	Hamra soil				Panoche soil			
S	$\bar{h}(S)$	CV	$\bar{k}(S)$	CV	$\bar{h}(S)$	CV	$\bar{k}(S)$	CV
0.750	24.8	3.9	0.138	0.92	15.1	0.67	0.015	2.46
0.775	13.2	3.3	0.166	0.86	13.4	0.67	0.021	2.27
0.800	7.7	2.8	0.198	0.80	11.6	0.67	0.030	2.02
0.825	4.8	2.2	0.236	0.74	9.8	0.65	0.040	1.73
0.850	3.3	1.6	0.283	0.68	8.2	0.63	0.056	1.48
0.875	2.4	1.2	0.339	0.61	6.6	0.59	0.080	1.27
0.900	1.9	0.81	0.410	0.53	5.0	0.50	0.120	1.07
0.925	1.5	0.52	0.498	0.45	3.7	0.39	0.185	0.95
0.950	1.3	0.30	0.614	0.34	2.4	0.27	0.282	0.74
0.975	1.1	0.13	0.772	0.20	1.2	0.06	0.451	0.54

†Mean value and coefficient of variation.

simulate classical Miller similitude (sometimes termed *macroscopic Miller similitude*):

$$a_h(S) \equiv h^*(S)/h(S) \qquad a_k^2(S) \equiv k(S)/k^*(S) \qquad [17]$$

where the asterisk denotes a field-wide mean value, calculated under the condition that the field-wide mean value of either a_h or a_k is 1.0 [see Jury et al. (1987) for the computational details]. The physical significance of Eq. [17] is that $h^*(S)$ and $k^*(S)$ represent the soil water properties of a field as a whole. They are components of a field-wide Richards equation having the form of Eq. [3b], with its field-wide water diffusivity given by Eq. [4] in terms of k^*, h^*, and S. Water movement in a specific field plot is characterized by a Richards equation that is transformed into the field-wide expression with the scaling relations in Eq. [7], but with a_h replacing a and a_k^2 replacing ω^2/μ in Eq. [8]. Thus, a_h and a_k function as Miller scale factors.

Table 2–3 summarizes the field-wide mean value and coefficients of variation for the scaling parameters a_h, a_k, and η (Eq. [14]) as computed by Jury et al. (1987) for the Hamra and Panoche soils. These three parameters are not independent, since Eq. [8], [13], and [14] lead to the relationship

$$a_h = a_k^{2/\eta} \qquad [18]$$

Equation [18], like Eq. [17], pertains to a given value of S. If $\eta = 2$, the two scale factors are equal. The results in Table 2–3 indicate that the relative variability of a_h and a_k is similar in the two soils (CV ≈ 0.45) and that η tends to be somewhat larger than 2, on the average. Jury et al. (1987) found that the distribution of η-values was highly skewed across each of the field data sets, but that Eq. [18] was obeyed nonetheless.

A detailed comparison of a_h and a_k with field data is a complicated process because of the need to separate the spatial variability of each parameter into deterministic ("drift") and random components (Jury et al., 1987). For both soils and both parameters, a linear "drift" term was

Table 2–3. Statistical analysis of the scale factors a_h and a_k.

Scale factor	Mean	CV	Corr. Model	C_n	C_o	a
			Hamra soil			m
a_h	1.000	0.435	E	0†	0.071	0.337
a_k	1.000	0.446	S	0	0.072	0.326
η	3.36	0.354	E	0	0.713	0.47
			Panoche soil			
a_h	1.000	0.445	S	0.106	0.023	15.12
a_k	1.000	0.477	S	0.088	0.113	3.32
η	2.54	0.471	E‡	0.129‡	0‡	0‡

†Data not provided in Jury et al. (1987).
‡Data for ln η, with mean 0.852 and variance 0.154.

adequate to express the nonrandom spatial variability. The random spatial variability then was epitomized by fitting the drift-free data to one of the two candidate covariance models:

$$\mathrm{cov}(l) = C_n\delta_{lo} + C_o \exp(-l/a) \qquad \text{(E-model)} \qquad [19]$$

$$\mathrm{cov}(l) = C_n\delta_{lo} + C_o \left[1 - \frac{3}{2}\left(\frac{l}{a}\right) + \frac{1}{2}\left(\frac{l}{a}\right)^3\right] \qquad \text{(S-model)} \qquad [20]$$

where C_n is the "nugget" variance contribution, C_o is the random variance, a is a range parameter, and l is the "lag distance" ($\delta_{lo} = 1$ if $l = 0$ and vanishes otherwise). If $C_n + C_o$ (the variance) and a have similar values for a_h and a_k, the two parameters may be judged statistically similar. Table 2–3 (last four columns) shows that this is indeed the case in the Hamra soil, but not so in the Panoche soil.

CONCLUDING REMARKS

Classical Miller similitude can be generalized to admit the invariance of the pore-size distribution, instead of the porosity, with the result that the first row in Table 2–1 becomes $S_* = S$ instead of $\theta_* = \theta$. This generalization is tantamount to the postulate that the characteristic length scale λ in Miller similitude is to be associated only with the pore space, not the porous medium as a whole. A concept of this kind recently has been exploited in microscopic theories of fluid flow through porous media (Johnson et al., 1986, 1987).

Generalized scaling analysis can be incorporated into the Richards equation by casting it first into dimensionless form with Eq. [5] and then into a scaled form with Eq. [7]. The latter transformation turns out to be possible only if the dependence of the water diffusivity, hydraulic conductivity, and matric potential on the volumetric water content is either a power-law or an exponential type. Irrespective of which of these two mathematical forms holds true, the relationship between pairs of hydraulic conductivity/matric potential values in a field soil must be an inverse power-law (Eq. [14]). If these necessary conditions are not met, the Richards equation cannot be scaled and still remain invariant in form.

Scaling analysis of the Richards equation shows that there are three interrelated scaling factors, a_h, a_k, and η (Eq. [18]). The first scales the matric potential, the second scales the hydraulic conductivity, and the third scales ln h in terms of ln k (Eq. [14]). Field data on soil water properties can be analyzed to verify the relationship among the three factors (evidence that a scale-invariant Richards equation applies) and to verify the equality of a_h and a_k (evidence that generalized Miller similitude applies). If Eq. [18] is verified with $\eta = 2$, then the field soil is Miller-similar in the generalized sense.

ACKNOWLEDGMENT

Gratitude is expressed to Ms. Terri DeLuca for her excellent typing of the manuscript.

REFERENCES

Danielson, R.E., and P.L. Sutherland. 1986. Porosity. p. 443–461. *In* A. Klute (ed.) Methods of soil analysis. Part 1. 2nd ed. Agron. Monogr. 9. ASA and SSSA, Madison, WI.

Hillel, D. 1980. Infiltration and surface runoff. p. 21. *In* Applications of soil physics. Academic Press, New York.

Johnson, D.L., J. Koplik, and R. Dashen. 1987. Theory of dynamic permeability and tortuosity in fluid-saturated porous media. J. Fluid Mech. 176:379–402.

Johnson, D.L., J. Koplik, and L.M. Schwartz. 1986. New pore-size parameter characterizing transport in porous media. Phys. Rev. Lett. 57:2564–2567.

Jury, W.A., D. Russo, and G. Sposito. 1987. The spatial variability of water and solute transport properties in unsaturated soil. II. Scaling models of water transport. Hilgardia 55:33–56.

Miller, E.E. 1980. Similitude and scaling of soil-water phenomena. p. 300–318. *In* D. Hillel (ed.) Applications of soil physics. Academic Press, New York.

Miller, E.E., and R.D. Miller. 1956. Physical theory for capillary flow phenomena. J. Appl. Phys. 27:324–332.

Nielsen, D.R., J.W. Biggar, and K.T. Erh. 1973. Spatial variability of field-measured soil-water properties. Hilgardia 42:215–260.

Russo, D., and E. Bresler. 1981. Soil hydraulic properties as stochastic processes: I. An analysis of field spatial variability. Soil Sci. Soc. Am. J. 45:682–687.

Simmons, C.S., D.R. Nielsen, and J.W. Biggar. 1979. Scaling of field-measured soil-water properties. Hilgardia 47:77–174.

Sposito, G. 1986. The 'physics' of soil water physics. Water Resour. Res. 22S:83–88.

Sposito, G. 1990. Lie group invariance of the Richards equation. p. 327–347. *In* J. Cushman (ed.) Dynamics of fluids in hierarchical porous media. Academic Press, New York.

Sposito, G., and W.A. Jury. 1985. Inspectional analysis in the theory of water flow through unsaturated soil. Soil Sci. Soc. Am. J. 49:791–798.

Sposito, G., and W.A. Jury. 1986. Group invariance and field-scale solute transport. Water Resour. Res. 22:1743–1748.

Tillotson, P.M., and D.R. Nielsen. 1984. Scale factors in soil science. Soil Sci. Soc. Am. J. 48:953–959.

Warrick, A.W., D.O. Lomen, and S.R. Yates. 1985. A generalized solution to infiltration. Soil Sci. Soc. Am. J. 49:34–38.

Warrick, A.W., and D.R. Nielsen. 1980. Spatial variability of soil physical properties in the field. p. 319–344. *In* D. Hillel (ed.) Applications of soil physics. Academic Press, New York.

Youngs, E.G., and R.I. Price. 1981. Scaling of infiltration behavior in dissimilar porous materials. Water Resour. Res. 17:1065–1070.

3 Application of Scaling to Soil-Water Movement Considering Hysteresis

E.G. Youngs

Cranfield Institute of Technology
Silsoe College
Silsoe, Bedford, England

The practical application of the physics of liquid behavior in porous materials concerns the physical state and flow through the bulk material. It does not specifically address the microscopic distribution of the liquid contained in the complex network of channels formed between the packed solid particles nor the complex velocity distribution in these irregularly shaped microchannels. Nevertheless, the bulk behavior, as obtained through measurements, is some average effect of that occurring in the micropores, so that fundamentally it is the behavior at the microscopic scale that determines the macroscopic behavior that is observed and measured.

The Millers' theory for capillary flow phenomena (Miller & Miller, 1955a,b; 1956) provides differential equations giving the macroscopic flow in porous materials from the assumption that the microscopic flow in the pore channels is governed by the physical laws of surface tension and viscous flow. In particular, it shows how reduced variables involving microscopic and macroscopic characteristic lengths permit scaling of Richards' equation for flow in similar unsaturated porous materials. Prominent in the discussion is that the theory recognizes the hysteretic nature of liquid retention and movement in porous media, being a function of the past history of wetting and draining of the pores.

The scaling of experimental results of soil-water behavior in similar porous materials is a consequence of the physical theory. Experiments performed on porous materials chosen so as to conform to being "similar" as near as possible, validate the basic assumptions that the Millers made in the development of theory. With the use of scaled variables relationships between physical quantities on different systems can be presented in a compact form, reducing the labor involved in calculations by relating characteristics of one system to those of another. It can also give an insight into physical factors affecting fluid behavior in porous bodies. Thus, the usefulness of scaling based on the Millers' theory would be enhanced if it could be extended to dissimilar porous materials. Youngs and Price (1981) showed this was possible for ponded infiltration relationships on a range of soil materials composed of particles of different shapes and sizes.

Theoretical studies on liquid movement in porous materials have mostly addressed situations where either wetting or draining separately were taking place throughout the flow region so that hysteresis did not have to be considered. Hysteretic situations in which both wetting and draining occur are commonplace, however. The most often cited one is the case of redistribution of soil water after the cessation of infiltration. Another situation where hysteretic conditions are produced is that of water uptake by bounded porous bodies where the increase in pressure of the displaced air has the effect of putting the water in a draining state after it has wetted the porous material. These two situations are used here as examples to show the usefulness of scaling based on similar media theory in the presentation of hysteretic soil-water movement in dissimilar porous materials, showing how diverse experimental results are placed in a more ordered form.

SCALED VARIABLES IN RICHARDS' EQUATION

Richards' (1931) equation describing water flow in unsaturated soils can be written as

$$\frac{\partial \theta}{\partial t} = \nabla \cdot (k \nabla p) + f \frac{\partial k}{\partial z} \qquad [1]$$

where θ, p, and k are the soil-water content, the soil-water pressure, and the capillary conductivity (equal to $K/\rho g$ where K is the hydraulic conductivity, ρ is the density of water, and g is the acceleration due to gravity) at a point (x, y, z) at a time t, and f is the body force per unit volume, equal to ρg for gravity. From the assumption that the microscopic behavior of a liquid in an unsaturated porous material is controlled by the physical laws of surface tension and viscous flow, for similar porous materials the Millers' theory allows Eq. [1] to be rewritten as

$$\frac{\partial \Theta}{\partial t^*} = \nabla^* \cdot (k^* \nabla^* p^*) + f^* \frac{\partial k^*}{\partial z^*} \qquad [2]$$

using scaled variables, defined in terms of the surface tension σ and viscosity η of the soil water, and a microscopic characteristic length λ of the porous material and a macroscopic characteristic length L of the flow region, by

$$k^* = (\eta/\lambda^2)k \quad [3a]$$

$$p^* = (\lambda/\sigma)p \quad [3b]$$

$$t^* = [\sigma\lambda/\eta L^2(\theta_0 - \theta_1)]t \quad [3c]$$

$$z^* = (1/L)z \quad [3d]$$

$$\nabla^* = L\nabla \quad [3e]$$

$$f^* = (\lambda L/\sigma)f \quad [3f]$$

where θ_0 and θ_1 are the saturated and initial soil-water contents, respectively, and in Eq. [2] $\Theta = (\theta - \theta_1)/(\theta_0 - \theta_1)$.

The microscopic characteristic length λ of a porous material can be any characteristic length of the porous matrix, for example, a characteristic pore size or a characteristic particle dimension. The macroscopic characteristic length L is a characteristic length of the flow region, for example the radius of an infiltrometer ring or the height of a draining soil column. The reduced variables defined by Eq. [3a] to [3f] show that both characteristic lengths are involved in scaling soil-water behavior, which is thus dependent not only on the soil but also on the dimensions of the flow region.

DEFINING THE MICROSCOPIC CHARACTERISTIC LENGTH

A convenient length that characterizes the geometry of the porous material has to be chosen for the microscopic characteristic length λ that occurs in the reduced variables of the scaled Richards' equation. Alternatively, the scaled hydraulic properties, such as those defined by Eq. [3a] and [3b], provide an indirect and sometimes more convenient means of obtaining a value. Thus, a microscopic characteristic length can be obtained from the hydraulic conductivity K_0 of the saturated porous material using the relationship

$$\lambda = \surd(\eta K_0/\rho g) \quad [4]$$

Additionally, defining λ indirectly in this way provides an easy means of extending the use of scaled variables, obtained using similar media theory, to soil-water behavior in dissimilar porous materials. Because in practice it would be very difficult for two porous materials to be exactly geometrically similar, such an extension is necessary if the theory is to have any general application.

SCALING INFILTRATION BEHAVIOR

Youngs and Price (1981) scaled the one-dimensional vertical infiltration into a range of soil materials with particles of different shapes and sizes using a microscopic characteristic length defined by Eq. [4]. Because the soil-water movement takes place in an effectively semi-infinite column of soil, there is no macroscopic characteristic length that characterizes the flow region. However, effectively Youngs and Price use a macroscopic characteristic length L given by

$$L = \sigma/\rho g\lambda \qquad [5]$$

that can be identified with the soil-water pressure head at which the porous material drains or wets up and makes $f^* = 1$. Figure 3–1 shows how the cumulative infiltration i into soil materials scales using the reduced variables

$$i^* = [1/L(\theta_0 - \theta_1)]i = [\rho g\lambda/\sigma(\theta_0 - \theta_1)]i \qquad [6a]$$

$$t^* = [\sigma\lambda/\eta L^2(\theta_0 - \theta_1)]t = [(\rho g)^2\lambda^3/\eta\sigma(\theta_0 - \theta_1)]t \qquad [6b]$$

The reduction of the diverse infiltration relationships shown in Fig. 3–1a to the single scaled relationship shown in Fig. 3–1b that fits the quasi-theoretical equation

$$i^* = 0.233\sqrt{t^*} + t^* \qquad [7]$$

that is a scaled form of Philip's (1957) infiltration equation, supports the use of scaled variables based on similar media theory in describing soil-water behavior in dissimilar soil materials with a microscopic characteristic length defined indirectly through the hydraulic conductivity of the saturated soil. Similarly, it was found (Youngs, 1982) that the infiltration and runoff behavior during constant rate infiltration could be scaled using the reduced variables given by Eq. [3a] to [3f]. Figure 3–2 shows the good agreement for different soil materials and a range of constant surface flow rates F_0 between the scaled time of ponding t_P^* and the scaled constant flux F_0^* given by

$$F_0^* = (\eta L/\sigma\lambda)F_0 \qquad [8]$$

which equals F_0/K_0 for the microscopic characteristic length defined by Eq. [4] and the macroscopic characteristic length defined by Eq. [5]. The scaled results are in good agreement with the quasi-theoretical equation

$$t_P^* = 0.027/[F_0^*(F_0^* - 1)] \qquad [9]$$

For infiltration from finite surface areas where the soil-water flow is

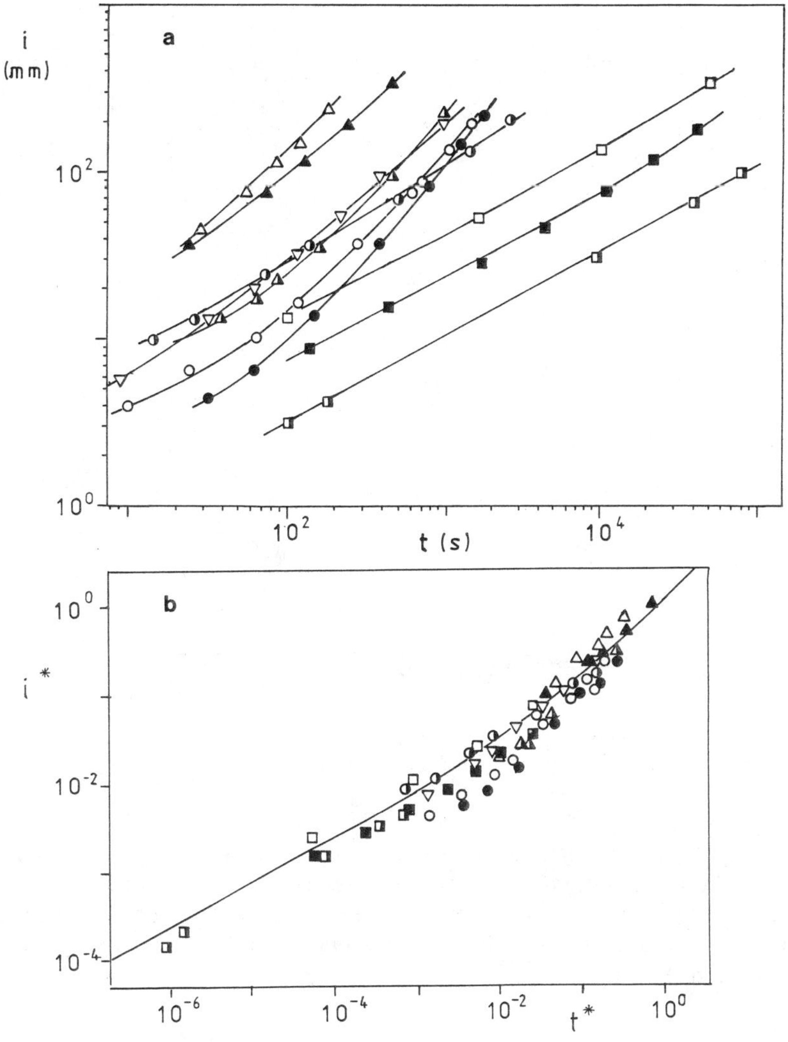

Fig. 3–1. (*a*) The cumulative infiltration *i* plotted against the time *t* for surface ponded infiltration for various porous materials: circles, glass beads (open, 210–325 μm particle size; closed, 115–180 μm; half closed, 60–95 μm); triangles, sands (open, Leighton Buzzard 350–500 μm; closed, Leighton Buzzard 250–350 μm; half closed, Leighton Buzzard, 180–250 μm; inverted open, graded beach sand, 180–250 μm); squares, packed soil materials (open, slate dust 40–125 μm; closed, Woburn sandy loam, mostly 60–350 μm; half closed, Rothamsted silt loam, mostly 2–60 μm).
(*b*) The scaled infiltration results plotted in terms of the reduced variables i^* anf t^* given by Eq. [6a] and [6b]. The line is Eq. [7].

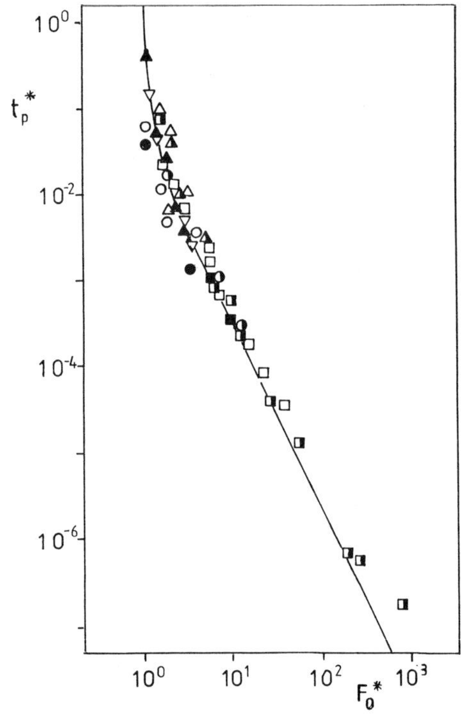

Fig. 3–2. The scaled ponding time t_p^* for constant flux infiltration plotted against the scaled constant flux F_0^*; (symbols as in Fig. 3–1). The line is Eq. [9].

no longer one-dimensional, the macroscopic characteristic length has to reflect the finite dimensions of the flow region. For an infiltration ring we can define it as the radius R, making the scaled body force f^* dependent on R. Thus, the relationship between the scaled total cumulative infiltration I^* through the ring, given by

$$I^* = I/R^3(\theta_0 - \theta_1) \qquad [10]$$

and the scaled time t^* given by Eq. [3c] depends on the radius of the ring. This is shown in the results given in Fig. 3–3 where the scaled values depart from a single relationship as t^* becomes large, although for small values of t^* ($t^* < {\sim}1.0$, $I^* < {\sim}1.5$) the results merge to a single relationship that is described by (Youngs, 1987)

$$I^* = 0.73\sqrt{t^*} + t^* \qquad [11]$$

Thus, for a ring of radius 100 mm, this relationship holds for water penetration depths up to about 50 mm. It is seen that I^* departs from being proportional to $\sqrt{t^*}$ at a very small value of t^* ($t^* \simeq 0.5$, $I^* \simeq 0.15$). Thus, for a ring of radius 100 mm, the square root of time behavior for the

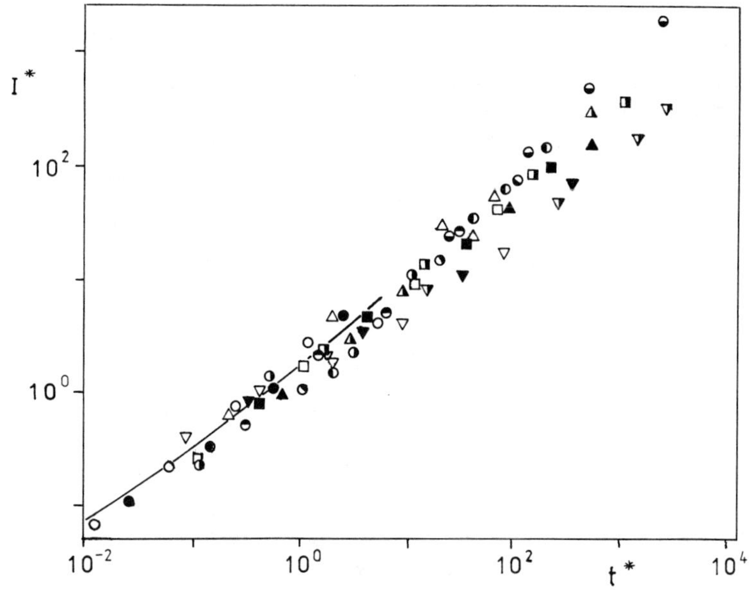

Fig. 3–3. The scaled total cumulative infiltration I^* given by Eq. [10] plotted against the scaled time t^* given by Eq. [3c] with $L = R$ for infiltrometer rings for various sands and soils: circles, Leighton Buzzard sand, particle size <2 mm; squares, silica sand, 180–250 μm; triangles, silty loam soil; inverted triangles, silty clay loam. The different shading of symbols show different infiltrometer ring radii that ranged between 455 mm and 10 mm for the case of the Leighton Buzzard sand and 49 mm and 12.6 mm for the other cases. The line is Eq. [11].

infiltration only occurs for infiltration penetration less than ~5 mm. This is in keeping with the experimental observation that the constant infiltration rate obtained with ring infiltrometers occurs after very short times, especially with small ring radii.

SCALING SOIL-WATER BEHAVIOR DURING THE REDISTRIBUTION OF INFILTRATION WATER

The infiltration process for which results were satisfactorily scaled using reduced variables, founded on similar media theory, is one in which the soil is always in a wetting state. In situations where soils wet and drain, the state of the soil water at a given time is dependent on the past history of wetting and draining because of the hysteretic relationships between soil-water content, soil-water pressure, and hydraulic conductivity. The classical situation in which this occurs is that of the redistribution of soil water after the cessation of infiltration.

After the cessation of infiltration, the soil water starts redistributing, with an advancing soil-water front continuing to wet up the soil at the expense of water draining from the region near the soil surface wetted during the infiltration. It has been found that the soil-water profiles can develop in two different ways, which Childs (1969) described as "a subject of considerable conflict of experimental evidence." Youngs and Poulovassilis (1976), however, showed that the different forms of profile development were consistent with classical soil-water theory that recognized hysteretic relationships in soil-water properties. The two forms are illustrated in Fig. 3–4. In the first (Fig. 3–4a), the soil-water profiles during redistribution maintain approximately the general shape of the infiltration profile with the soil-water content gradient everywhere positive and with the soil-water content almost uniform above a sharp wetting front (Shaw, 1927; Biswas et al., 1966; Staple, 1966, 1969; Gardner et al., 1970). In the second type of behavior (Fig. 3–4b) the wetting front at the cessation of infiltration persists with further wetting below advancing as a step-like profile while the soil desaturates near the surface (Youngs, 1958b; Biswas et al., 1966; Childs, 1969; Talsma, 1974). Although Youngs and Poulovassilis (1976) were able to explain why these two types of profile occurred, quantitative predictions have to be made through numerical calculations requiring extensive data for the hysteretic soil-water relationships. Scaling based on similar media theory provides an alternative method of obtaining an estimate of the soil-water redistribution.

The depth of the wetting front at the cessation of infiltration z_0 determines the subsequent moisture profile development for a given soil during the hysteretic redistribution process and can be conveniently taken as the macroscopic characteristic length in the scaled variables of Eq. [2] that describes the soil-water movement. Because $z_0^* = 1$ for all soil materials, the $\Theta - z^*$ profiles at the cessation of infiltration are practically

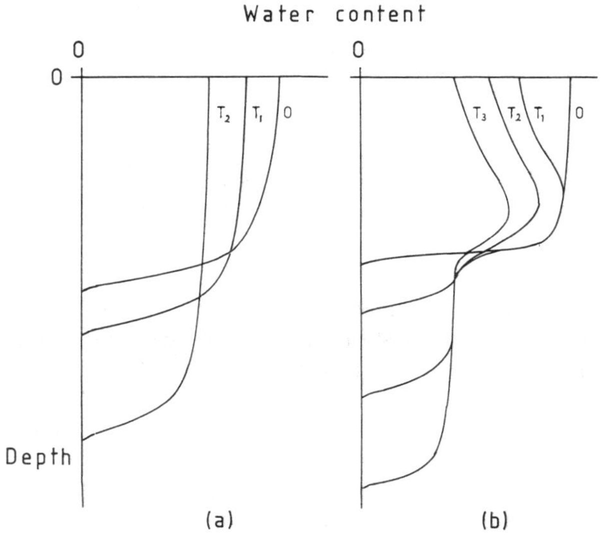

Fig. 3–4. The two different forms of moisture profile development after the cessation of infiltration.

the same in all cases, so that the subsequent redistribution soil-water profiles are described by some function

$$\Theta = \Theta(z^*, T^*, f^*) \qquad [12]$$

where T^* is the scaled time from the start of redistribution. The development of the profiles is thus dependent on the scaled body force $f^* = (\lambda z_0/\sigma)\rho g$ and hence on the physical properties of the soil material through λ and the depth of the final infiltration profile z_0, as concluded from the theoretical considerations of Youngs and Poulovassilis (1976). Scaled redistribution profiles can thus be shown as a family of plots of Θ against z^* for a range of T^* values and a range of f^*. Alternatively, as done in a previous discussion on scaling of the redistribution process (Youngs, 1983), the progress of redistribution can be followed by considering the average soil-water content $\bar{\theta}$. The scaled time $T^*_{\bar{\Theta}}$ at which the average normalised soil-water content has been reduced to $\bar{\Theta}$ can then be expressed by some function

$$T^*_{\bar{\Theta}} = T^*_{\bar{\Theta}}(f^*) \qquad [13]$$

In Fig. 3–5a is shown the times $T_{0.5}$ at which the average soil-water content was reduced to 50% of the initial average soil-water content of the infiltration profile for different initial infiltration depths z_0 in different soil materials in redistribution experiments reported by various workers. It is seen that there is no order in this presentation of results. However, when the reduced time $T^*_{0.5}$ at which $\bar{\Theta} = 0.5$ is plotted against f^* as in Fig. 3–5b, the results become ordered into some functional relationship as is suggested from similar media theory. Also shown in Fig. 3–5b are two results for redistribution profiles in horizontal columns: these are equivalent to results for $z_0 \rightarrow 0$.

In the plots of Fig. 3–5a and 3–5b showing the results of redistribution experiments, the different forms of redistribution profile as depicted in Fig. 3–4a and 3–4b are differentiated by underlining the results for the redistribution developing in the form shown in Fig. 3–4b, and not underlining those developing in the form of Fig. 3–4a. It is seen that, when the results are presented in reduced variables as in Fig. 3–5b, the results for redistribution of the form Fig. 3–4a all lie together at small values of f^* corresponding to large values of T^*, whereas those for redistribution of the form Fig. 3–4b lie together at larger values of f^* with correspondingly smaller values of T^*, with the change-over occurring at $f^* \simeq 0.05$. This is in keeping with the theoretical conclusions of Youngs and Poulovassilis (1976) because the form of Fig. 3–4a occurs with finer textured soils (small λ) and small infiltration depths (small z_0), and the form of Fig. 3–4b occurs with coarser-textured soils (large λ) and large infiltration depths (large z_0), as reflected in the values of f^*

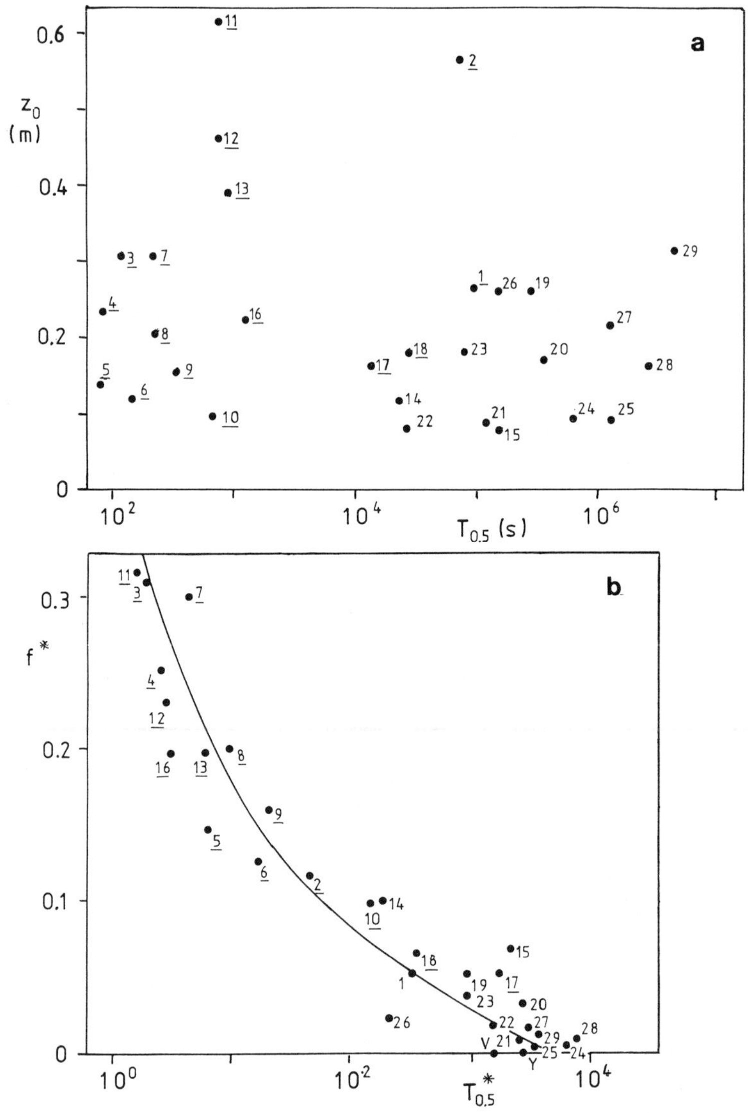

Fig. 3–5. (*a*) The times $T_{0.5}$ after the cessation of infiltration at which the average soil-water content had been reduced to 0.5 of saturation, shown plotted against the infiltration depth for various experiments: 1, 2 slate dust (Youngs, 1958b); 3–6 Leighton Buzzard sand 350–500 μm, 7–10 Leighton Buzzard sand 250–350 μm, 10–13 graded beach sand 180–250 μm; 14–16 Spearwood sand, 17, 18 Bungendore fine sand, 19–20 Glebe loam (Talsma, 1974); 21 Rideau clay, 22, 23 Upland sand, 24 Castor loam (Staple, 1969); 25 Rothamsted silt loam; 26, 27 Woburn sandy loam; 28–29 Gilate loess fine sandy loam (Gardner et al., 1970). The underlined numbers indicate that redistribution was of the form shown in Fig. 3–4b while numbers that are not underlined indicate that the redistribution was of the form depicted in Fig. 3–4a.
(*b*) The scaled times $T^*_{0.5}$ given by Eq. [3c] with $L = z_0$ plotted against the scaled body force f^* given by Eq. [3f]. The points *V* and *Y* were obtained from results for the redistribution after horizontal infiltration into Mont Cenis silt loam (Vachaud, 1966) and slate dust (Youngs, 1958a), respectively.

SCALING WATER UPTAKE BY BOUNDED POROUS BODIES

Infiltration theory generally assumes that the air displaced by the infiltrating water has free access to the atmosphere so that there is no build-up of air pressure in front of the wetting front that would impede the flow of water. In situations where the air has no means of escape, however, as occurs when the soil surface is completely flooded or when peds are surrounded by macropores full of water, the pressure of the air inside the soil rises as it is compressed by the water being imbibed by the soil. This gives rise to the soil water in the pores changing from being in a wetting state to being in a draining state (Youngs & Peck, 1964). When the pressure rises to the air-entry value of the soil-water pressure relationship, air starts bubbling through the surface. Peck (1965a,b) gave results of experiments showing how the air pressure build-up and final escape varied with the soil material and dimensions of the bounded porous body.

The water uptake by a bounded porous material is described by Richards' equation (Eq. [1]) subject to the boundary condition imposed by the increasing pressure of the compressed air. This air pressure acts on the water menisci in the soil pores, so that the effective soil-water pressure that relates to the soil-water content is the atmospheric pressure less this pressure; that is, the boundary conditions for flow can be taken as the same as applying a changing negative soil-water pressure at the surface infiltrating into an unimpeded body.

The process of water uptake by bounded porous bodies can be characterized by the penetration distance of the wetting front when air starts escaping through the surface. For one-dimensional flow, applying Boyle's Law as did Youngs and Peck (1964), this distance L_E is

$$L_E = L_B p_E/(p_E + A) \simeq L_B p_E/A \qquad [14]$$

where L_B is the length of the column, p_E the excess air pressure at which air starts escaping through the infiltrating surface, and A is the atmospheric air pressure and generally $p_E \ll A$. Now p_E may be assumed to be given in terms of the microscopic characteristic length by

$$p_E = 2\sigma/\lambda \qquad [15]$$

so that

$$L_E = 2\sigma L_B/A\lambda \qquad [16]$$

Thus, the scaled cumulative water uptake per unit area i^* and scaled time t^* can be defined for an initially dry column by

$$i^* = (A\lambda/2\eta L_B\theta_0)i \qquad [17a]$$

$$t^* = (A^2\lambda^3/4\eta L_B^2\sigma\theta_0)t \qquad [17b]$$

where i is the actual water uptake per unit area at a time t after the start of the wetting-up process.

Peck (1965a) reported results of experiments on the water uptake of horizontal bounded columns of porous materials. The water uptake per unit area as a function of the time t for different column lengths of a slate dust and a sand are shown in Fig. 3–6a. These results merge into a single relationship as shown in Fig. 3–6b when they are scaled using the reduced variables defined in Eq. [17a] and [17b].

DISCUSSION

The use of reduced variables in scaling physical behavior is made use of in all branches of engineering. One example is the use of nondimensional reduced variables in land drainage where water-table heights expressed as a fraction of the drain spacing are given in terms of the rainfall rate expressed as a fraction of the hydraulic conductivity of the saturated soil, making it unnecessary to treat each drain spacing and soil separately. Such scaling evolves from macroscopic considerations of the flow only and does not extend to considerations of the unsaturated soil-water zone. To do this appeal has to be made to the microscopic behavior of the soil water in the pore channels, as was done in the Millers' similar media theory. It is seen from this theory that scaled soil-water flows are dependent on both the microscopic characteristic length that determines unsaturated soil-water properties and the macroscopic characteristic length that describes the flow region.

The Millers' theory is based on the consideration of viscous fluid movement and surface tension forces in geometrically similar porous materials. Early experiments with sand fractions that were chosen to provide geometrical similarity (Klute & Wilkinson, 1958; Wilkinson & Klute, 1959) validated the theory. The usefulness of the theory to the real world is in its application to a wide variety of soils whose microstructures are not geometrically similar. The results used as illustrations in this chapter show that the extension of the theory to dissimilar soils is possible.

In using similar media theory in modeling soil-water behavior in dissimilar soils, the problem arises as to the definition of a microscopic characteristic length. This has been solved by defining this length indirectly from hydrodynamic properties of the soil, such as the hydraulic conductivity, because the theory shows a relationship between these properties and the microscopic characteristic length. Using a microscopic characteristic length defined from the hydraulic conductivity of the saturated soil, one-dimensional infiltration would scale at large times because the infiltration rate is then the hydraulic conductivity. Similarly, if a microscopic characteristic length were obtained from the sorptivity, then scaling would naturally be good at small times. The application of similar media theory to dissimilar soils is then assisted by physical macroscopic laws.

The physics of hysteresis in soil-water properties that Haines (1930)

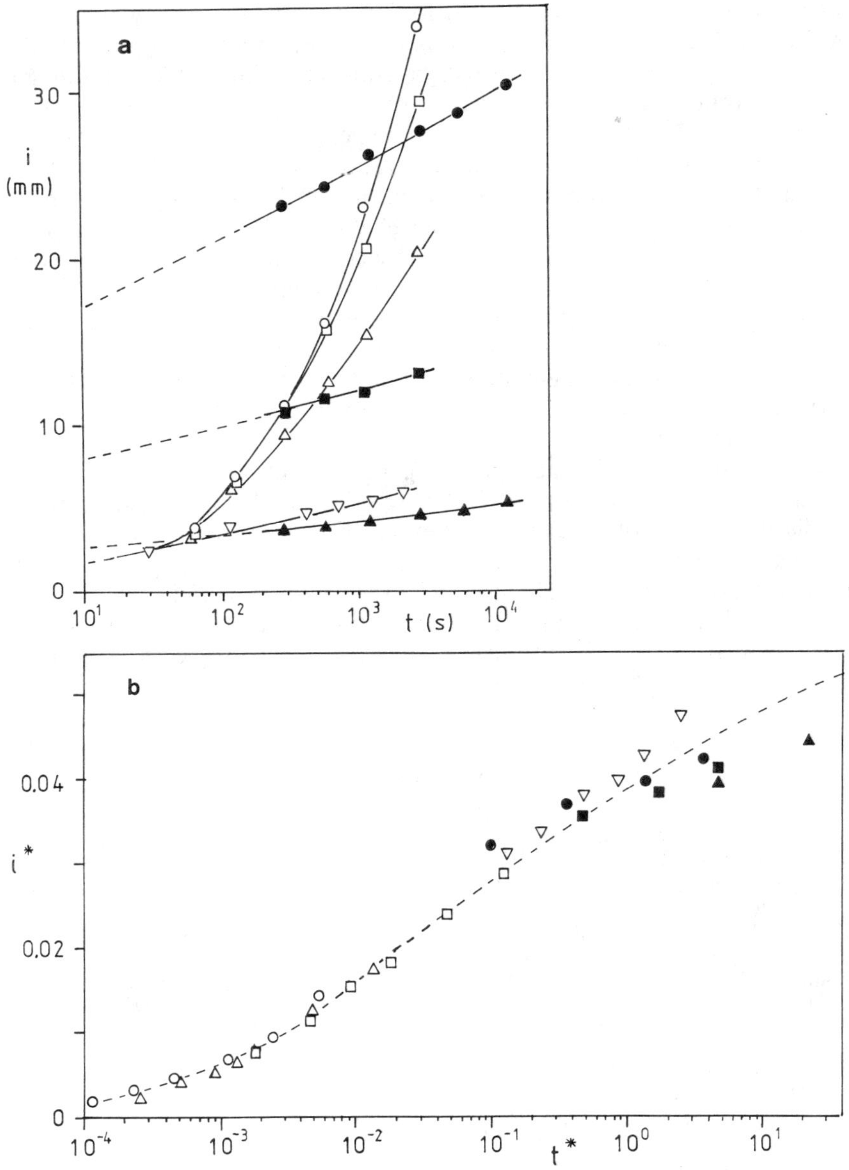

Fig. 3–6. (*a*) The cumulative water uptake i plotted against the time t for bounded columns of porous material: open symbols are for columns of slate dust of length 2.9 m (circles), 1.98 m (squares), 1.09 m (triangles), and 0.13 m (inverted triangles); closed symbols are for sand columns of length 5.3 m (circles), 2.4 m (squares) and 0.7 m (triangles). (*b*) The scaled water uptake i^* given by Eq. [17a] plotted against the scaled time t^* given by Eq. [17b].

discussed was an integral part of the Millers' theory of fluid flow in porous materials. The inclusion of hysteresis in the theory of soil-water profile development is not simple, however, as shown by Childs (1969), although this hysteretic theory shows that two forms of redistribution profile after the cessation of infiltration are possible, depending on the soil-water properties and the depth of infiltration (Youngs & Poulovassilis, 1976). Similar media theory allows the soil-water behavior during such hysteretic processes to be scaled, giving some order to the complex pattern of results. Thus, it is possible to predict the occurrence of one or the other forms of profile development during the redistribution process. It also allows a prediction of the slowing down of the water uptake by bounded porous bodies as air is compressed by the advancing wetting front and an estimation of the time when air might be expected to start bubbling through the surface. These hysteretic processes are largely ignored in discussions of soil-water movement.

Because of the dissimilar nature generally between one porous material and another, accurate predictions of soil-water behavior using similar media theory cannot be expected. Perhaps because of the very nature of porous materials with their inherent heterogeneity, however, precision can never be expected. What similar media theory offers is a means to place a scale on behavioral events, allowing the time scale for soil-water development to be obtained from microscopic and macroscopic characteristic lengths, knowing the development for one.

REFERENCES

Biswas, T.D., D.R. Nielsen, and J.W. Biggar. 1966. Redistribution of soil water after infiltration. Water Resour. Res. 2:513–524.

Childs, E.C. 1969. An introduction to the physical basis of soil water phenomena. John Wiley & Sons, New York.

Gardner, W.R., D. Hillel, and Y. Benyamini. 1970. Post-irrigation movement of soil water. 1. Redistribution. Water Resour. Res. 6:851–861.

Haines, W.B. 1930. Studies in the physical properties of soils: V. The hysteresis effect in capillary properties and the modes of moisture distribution associated therewith. J. Agric. Sci. 20:97–116.

Klute, A., and G.E. Wilkinson. 1958. Some tests of the similar media concept of capillary flow: I. Reduced capillary conductivity and moisture characteristic data. Soil Sci. Soc. Am. Proc. 22:278–280.

Miller, E.E., and R.D. Miller. 1955a. Theory of capillary flow: I. Practical implications. Soil Sci. Soc. Am. Proc. 19:267–271.

Miller, E.E., and R.D. Miller. 1956. Physical theory for capillary flow phenomena. J. Appl. Phys. 27:324–332.

Miller, R.D., and E.E. Miller. 1955b. Theory of capillary flow: II. Experimental information. Soil Sci. Soc. Am. Proc. 19:271–275.

Peck, A. J. 1965a. Moisture profile development and air compression during water uptake by bounded porous bodies: 2. Horizontal columns. Soil Sci. 99:327–334.

Peck, A. J. 1965b. Moisture profile development and air compression during water uptake by bounded porous bodies: 3. Vertical columns. Soil Sci. 100:44–51.

Philip, J.R. 1957. The theory of infiltration: 4. Sorptivity and algebraic infiltration equations. Soil Sci. 84:257–264.

Richards, L.A. 1931. Capillary conduction of liquids through porous mediums. Physics 1:318–333.

Shaw, C.F. 1927. The normal moisture capacity of soils. Soil Sci. 23:303–317.

Staple, W.J. 1966. Infiltration and redistribution of water in vertical columns of loam soil. Soil Sci. Soc. Am. Proc. 30:553–558.

Staple, W.J. 1969. Comparison of computed and measured moisture redistribution following infiltration. Soil Sci. Soc. Am. Proc. 33:840–847.

Talsma, T. 1974. The effect of initial moisture content and infiltration quantity on redistribution of soil water. Aust. J. Soil Res. 12:15–26.

Vachaud, G. 1966. Essai d'analyse de la redistribution après l'arret d'une infiltration dans une colonne horizontale de sol non saturé. C.R. Acad. Sci. 262:839–842.

Wilkinson, G.E., and A. Klute. 1959. Some tests of the similar media concept of capillary flow: II. Flow systems data. Soil Sci. Soc. Am. Proc. 23:434–437.

Youngs, E.G. 1958a. Redistribution of moisture in porous materials after infiltration: 1. Soil Sci. 86:117–125.

Youngs, E.G. 1958b. Redistribution of moisture in porous materials after infiltration: 2. Soil Sci. 86:202–207.

Youngs, E.G. 1982. Use of similar media theory in infiltration and runoff relationships. p. 149–162. *In* V.P. Singh (ed.) Proc. Int. Symp. Rainfall/Runoff Modeling, Mississippi State University, May 1981, Modeling Components of the Hydrologic Cycle. Water Resources Publications, Littleton, CO.

Youngs, E.G. 1983. The use of similar media theory in the consideration of soil-water redistribution in infiltrated soils. p. 48–54. *In* Proc. Conf. Advances in Infiltration, Chicago. December 1983. Am. Soc. Agric. Eng., St. Joseph, MI.

Youngs, E.G. 1987. Estimating hydraulic conductivity values from ring infiltrometer measurements. J. Soil Sci. 38:623–632.

Youngs, E.G., and A.J. Peck. 1964. Moisture profile development and air compression during water uptake by bounded porous bodies: 1. Theoretical introduction. Soil Sci. 98:290–294.

Youngs, E.G., and A. Poulovassilis. 1976. The different forms of moisture profile development during the redistribution of soil water after infiltration. Water Resour. Res. 12:1007–1012.

Youngs, E.G., and R.I. Price. 1981. Scaling of infiltration behavior in dissimilar porous materials. Water Resour. Res. 17:1065–1070.

4 Application of Scaling to the Characterization of Spatial Variability in Soils

A.W. Warrick

University of Arizona
Tucson, Arizona

Early applications of the theory of similar media were by Miller and Miller (1955a,b, 1956), which were later summarized by Miller (1980). Included was a rigorous analysis of the consequences of similar systems. In Fig. 4–1 are two soil-water systems differing only by scale factors λ_1 and λ_2. The single scale factor describes either system exactly relative to the other. In the original papers, Miller and Miller proved rigorously for a given water content the capillary pressures h and unsaturated hydraulic conductivity K for such similar systems are related:

$$\lambda_1 h_1 = \lambda_2 h_2 = \lambda_{avg} h_{avg} \quad [1]$$

$$K_1/\lambda_1^2 = K_2/\lambda_2^2 = K_{avg}/\lambda_{avg}^2 \quad [2]$$

What follows here is based on these two equations. The presentation is given by example with minimal references. The objective is to demonstrate the application of this theory to characterize spatial variability of soils. Examples will be given under three headings: data reduction, generalized solutions, and simulations.

APPLICATION EXAMPLES—DATA REDUCTION

Equations [1] and [2] are useful to reduce large volumes of data into average relationships using scaling factors to characterize individual sites. Thus, qualitatively we seek to coalesce data into meaningful averages while preserving variability through a set of site-specific factors.

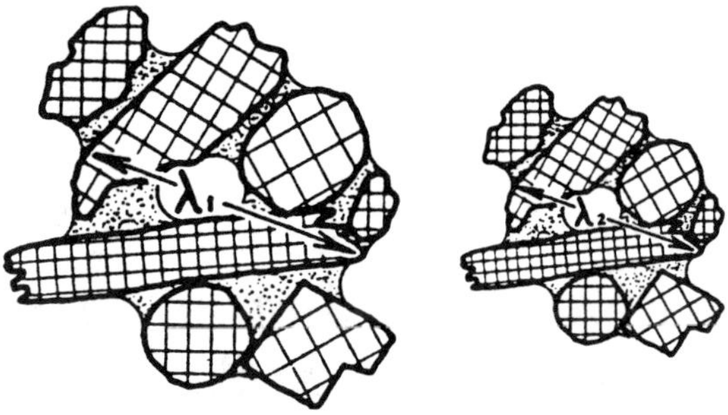

Fig. 4–1. Similar media (after Miller and Miller, 1956).

Example 1. Reduction of Soil Water Characterization and Unsaturated Hydraulic Conductivities for the Panoche 1. Warrick et al. (1977) considered the data of Nielsen et al. (1973) for the Panoche soil [fine-loamy, mixed (calcareous), thermic Typic Torriorthents] of California. Included were soil water characteristic curves based on results from six depths at 20 sites within a 150-ha area. For each characteristic curve seven corresponding h-θ values were determined. Additionally, 2640 unsaturated hydraulic conductivity points were available from data collected with time after the 20 sites were flooded and allowed to drain.

The assumptions were relaxed as follows:

1. Equations [1] and [2] were assumed to apply for equal degree of saturation $S = \theta/\theta_s$ (with θ_s the saturated water content).
2. The average h and K was assumed to be functions

$$h_{avg}(S) = a_i h(S) \qquad [3]$$

$$K_{avg}(S) = K(S)/a_i^*2 \qquad [4]$$

The first assumption forces water amounts between 0 and 1, rather than having a different maximum value for each sample. The original assumptions illustrated in Fig. 4–1 can never exactly be met. To have the same value for saturation the normalization (θ/θ_s) is performed. The second assumption requires choosing a functional form. This allows a best fit, because generally measurements are not for equal water content increments. An added advantage is that "average curves" for $h(S)$ and $K(S)$ are defined in the process. Also, Eq. [3] and [4] are consistent with relaxation of the original assumptions. Assumption 2 allows the scaling factors a and a^* to be different.

Values of a_i were best fit by minimizing the sum of squares

$$SS = \sum_{r,i} [\hat{h}_{r,i} - a_r h_{r,i}] \qquad [5]$$

where r is a sample index and $i = 1, \ldots, 7$ is a soil water pressure index.

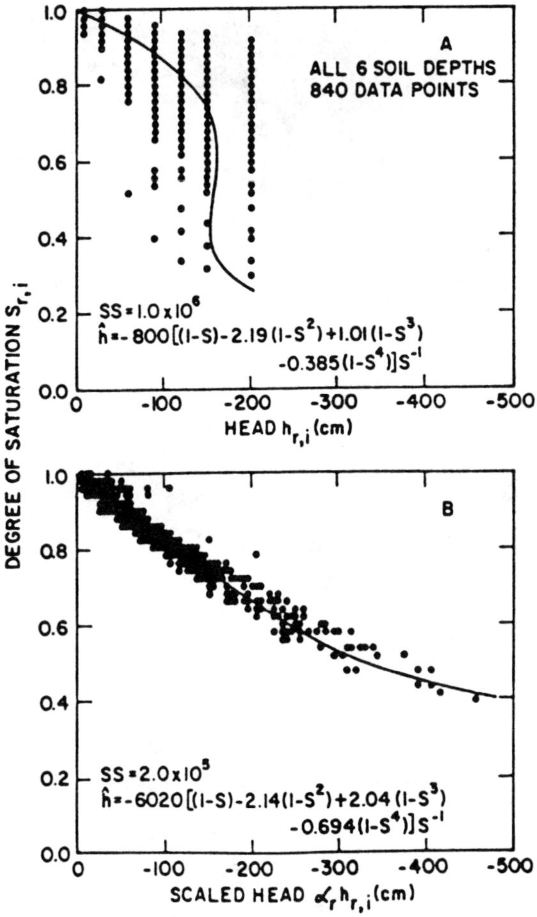

Fig. 4–2. Soil water characteristic data for Panoche soil: (*A*) unscaled and (*B*) scaled (after Warrick et al., 1977).

Results are presented as Fig. 4–2 for the combined 840 data points. Part A is the original data along with a fitted, power curve with four coefficients. Figure 4–2B is the reduced curve based on Eq. [3] after choosing 120 values of a and four coefficients for h based on Eq. [5]. The result is an 80% reduction in the SS between the fitted function of Part A to Part B. Individual curves tend to move toward the central or averaged curve.

For the 2640 unsaturated hydraulic conductivity values, the sum of squares

$$SS = \sum_{i,r} (\hat{\ln} K_{r,i} + 2 \ln a_r^* - \ln K_{r,i})^2 \quad [6]$$

was minimized with results given in Fig. 4–3. Average curves are third-order polynomials for ln K. A reduction in sum of squares occurs of about

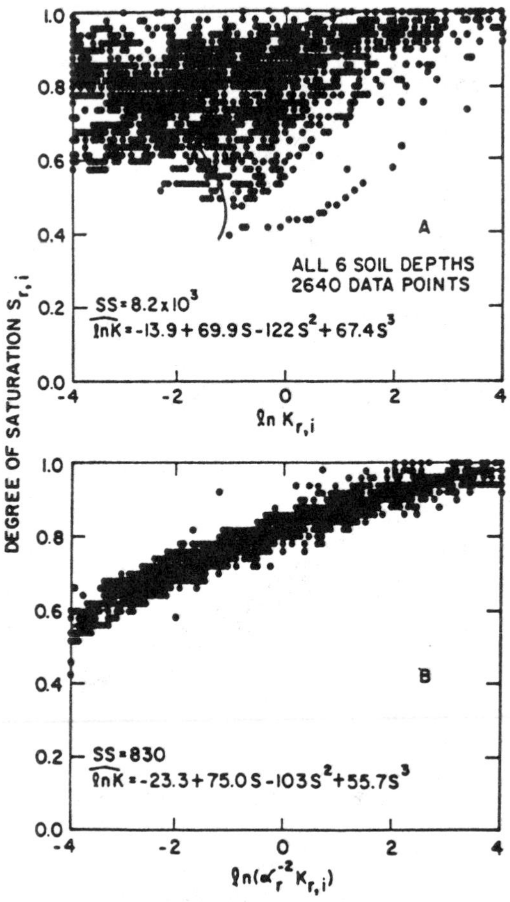

Fig. 4–3. Hydraulic conductivity data for Panoche soil: (*A*) unscaled and (*B*) scaled (after Warrick et al., 1977).

an order of magnitude using 120 values of a^* (one for each site and depth combination).

The values of scaling coefficients a^* tend to be larger than a for the same sites as illustrated in a scattergram (Fig. 4–4), although they have a correlation of $r = 0.9$. In other studies, Russo and Bresler (1980) found a higher correlation but Rao et al. (1983) found essentially zero correlation for the Lakeland fine sand. Sposito and Jury (1985) point out that there is no reason, a priori, for a and a^* to be correlated for real systems.

Example 2. Scaling Techniques to Quantify Variability in Hydraulic Functions of Soils in the Netherlands. Wösten (1989) used scaling techniques to quantify variability in hydraulic functions in the Netherlands. A total of 197 water retention and hydraulic conductivity functions were measured for different soils by a combination of methods. The soil char-

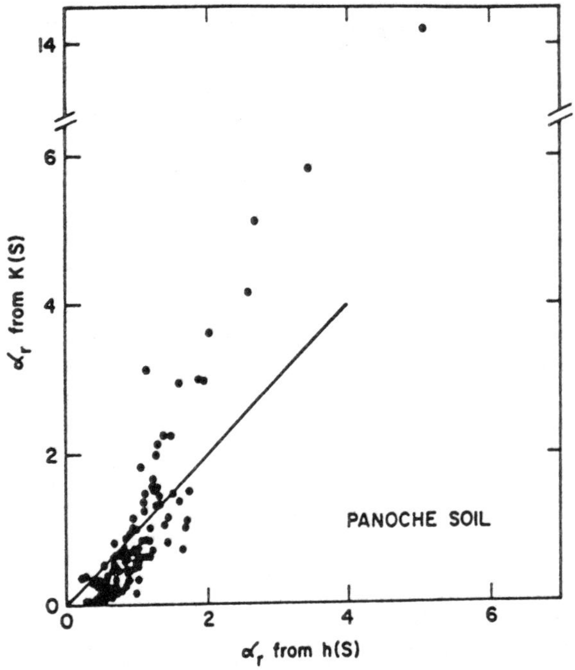

Fig. 4–4. Scaling parameters calculated from hydraulic conductivity data vs. those calculated from soil water characteristic data for Panoche soil (after Warrick et al., 1977).

acteristic curves for coarse-textured soils are in Fig. 4–5A and the unsaturated conductivity curves in Fig. 4–5C. The coarse-textured group included 105 of the 197 functions. Also shown as Fig. 4–5B and D are the reduced soil water characteristics and unsaturated hydraulic conductivity after scaling. The reduction in sum of squares before and after was 80 and 53% for h and K, respectively.

Example 3. The Las Cruces Trench. In New Mexico, 450 cores were collected from a 25 m by 6 m deep trench wall (50 samples from each of nine identified soil layers with a horizontal spacing of 0.5 m) (Hills et al., 1989). Water contents were measured at 11 tension values for each core (Fig. 4–6A). Scaling was performed with a based on van Genuchten's (1980) relationship:

$$\Theta = \frac{\theta - \theta_r}{\theta_{\mathrm{sat}} - \theta_r} = [1 + |ah|^n]^{(1-n)/n} \qquad [7]$$

where θ_r was chosen as the water retained at 1.5 MPa tension and θ_{sat} was chosen as the measured saturated water content. A single value of n was best fit ($n = 1.53$) along with 450 values of a to give the results shown in Fig. 4–6B. The reduction in mean square error between predicted and measured values is about 80%.

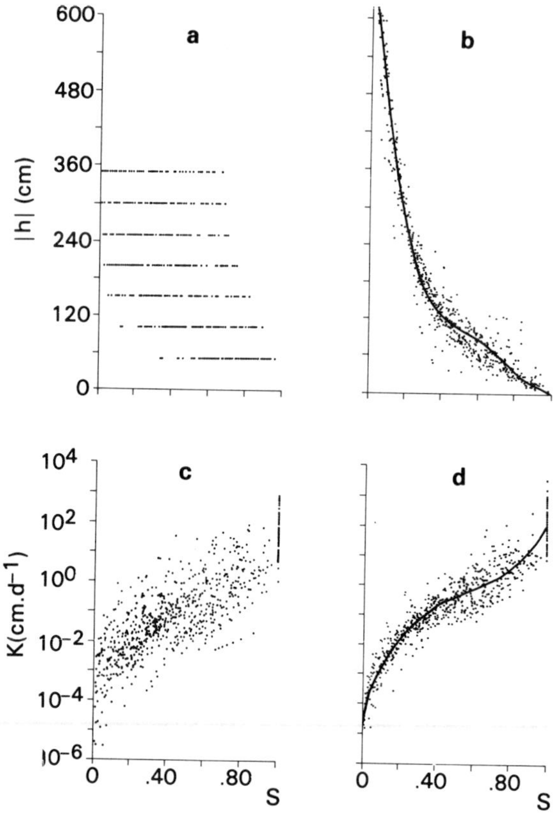

Fig. 4–5. Soil water characteristic curves and hydraulic conductivity for 105 coarse-textured Dutch soils (after Wösten, 1989).

APPLICATION EXAMPLES—GENERALIZED SOLUTIONS

Miller (1980) discusses "telescoping of flow system equations" to apply single solutions to a multiplicity of boundary-value problems. This was done by Warrick et al. (1985) for infiltration. The Richards Equation is considered in reduced form:

$$\partial S/\partial T = (\partial/\partial X)\,(D^*\partial S/\partial X) - \partial K^*/\partial X \qquad [8]$$

where S, T, X, D^*, and K^* are from

$$S = (\theta - \theta_r)/(\theta_s - \theta_r) \qquad [9]$$

$$T = aK_s t/(\theta_s - \theta_r) \qquad [10]$$

$$X = ax \qquad [11]$$

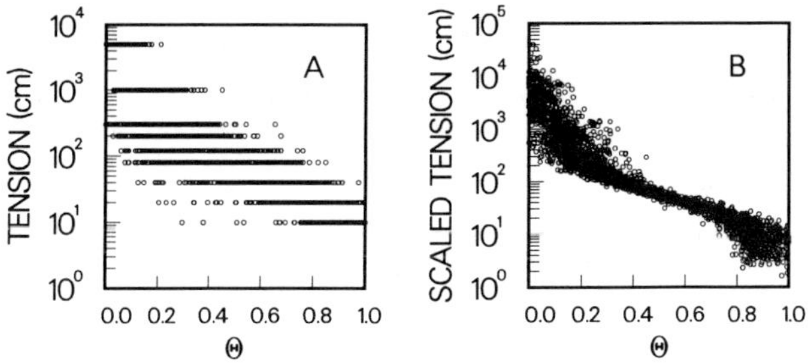

Fig. 4–6. Soil water characteristic curves before and after scaling (after Hills et al., 1989).

$$K^* = K/K_s \quad [12]$$

$$h^* = \alpha h \quad [13]$$

$$D^* = K^* dh^*/dS = \alpha D(\theta_s - \theta_r)/K_s. \quad [14]$$

The θ_s and θ_r are saturated and residual water contents. Equation [8] is applied for any form of water content for which K and S can be expressed as functions of $h^* = \alpha h$. This includes all relationships that satisfy Eq. [3] and [4], including van Genuchten's (1980) relationship of Eq. [7] and

$$K^* = S^{0.5}[1 - (1 - S^{n/(n-1)})^{1-1/n}]^2 \quad [15]$$

If initial and boundary conditions are for infiltration:

$$S(X, 0) = S_i \quad [16]$$

$$S(0, T) = S_f \quad [17]$$

$$\lim_{X \to \infty} S(X, T) = S_i \quad [18]$$

then Philip's (1969) quasi-analytical solution can be written as

$$X = \lambda(S)T^{1/2} + \chi(S)T + \psi(S)T^{3/2} \ldots \quad [19]$$

where $\lambda(S)$, $\chi(S)$, and $\psi(S)$ and any additional terms are for the dimensionless parameters. A dimensionless form for the cumulative infiltration is

$$I^* \approx AT^{1/2} + BT + CT^{3/2} \quad [20]$$

with

$$A = \int_{S_i}^{S_f} \lambda(S)\, dS \quad [21]$$

Table 4–1. Infiltration coefficients for Example 4 with $n = 2$ (after Warrick et al., 1985).

W_i	A	B	C	T_g
0.0	0.883	0.367	0.119	0.780
0.1	0.836	0.370	0.126	0.699
0.2	0.786	0.374	0.134	0.618
0.3	0.733	0.378	0.143	0.538

$$B = \int_{S_i}^{S_f} \chi(S)\, dS + K(S_i)/K_s \qquad [22]$$

$$C = \int_{S_i}^{S_f} \psi(S)\, dS \qquad [23]$$

The surface intake rate is

$$v_0 = dI/dt \qquad [24]$$

or

$$v_0 = K_s v_0^* \qquad [25]$$

Example 4. Generalized Results for Infiltration. Values for A, B, and C are presented by Warrick et al. (1985) for six representative values of "n" (from Eq. [7] and [15]) and for initial water contents corresponding to $S_i = 0, 0.1, 0.2$, and 0.3. Values are repeated in Table 4–1 for $n = 2$. For the Yolo light clay (fine-silty, mixed, nonacid, thermic Typic Xerorthents) (Moore, 1938) values of $n = 2$, $a = 1.5\ \mathrm{m}^{-1}$, $\theta_s = 0.495$, $\theta_r = 0.24$, and $K_s = 1.23(10)^{-7}$ m/s were chosen. By Eq. [25] the infiltration rate may be plotted as in Fig. 4–7. The value T_g in Table 4–1 corresponds to the large-time cutoff value of Philip (1969, p. 250), which for this example is about 300 h.

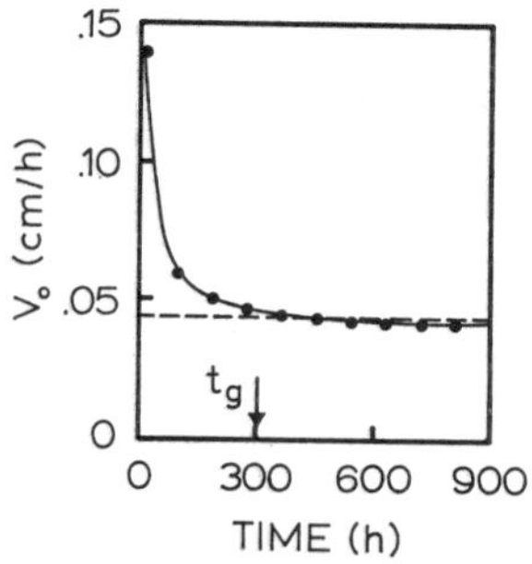

Fig. 4–7. Infiltration curve for Yolo light clay (after Warrick et al., 1985).

The values of A, B, and C used for Table 4–1 are valid for all α and K_s, which follow van Genuchten's (1980) form. They are restricted only by the choice of n and S_i. Further examples are in Warrick et al. (1985).

APPLICATION EXAMPLES—SIMULATIONS

Example 5. Effects of Spatial Variability in Water Budget Modeling. Peck et al. (1977) used the relationships of Eq. [1] and [2] to simulate water balance on a forest in eastern Tennessee. This was based on measured hydraulic conductivities and soil water characteristics of the Fullerton soil (clayey, kaolinitic, thermic, Typic Paleudults). The water budget model PROSPER was used to combine meteorological, soil, and plant systems.

A sample output for April is in Fig. 4–8 showing a partitioning of interception, evaporation, drainage, and gain in soil storage as a function of the scaling factor α (equivalent to Eq. [3] and α^* of Eq. [4]). If the factors are the same, larger α values correspond to coarser-texture soils with more drainage and less storage.

Example 6. Stochastic Analysis of Soil Water Regime. Hopmans and Sticker (1989) conducted a stochastic-deterministic analysis with respect to the Hupselse Beek Watershed in the eastern part of the Netherlands. The first objective was to describe measured variation of model parameters and input parameter describing the soil water regime. The ultimate goals

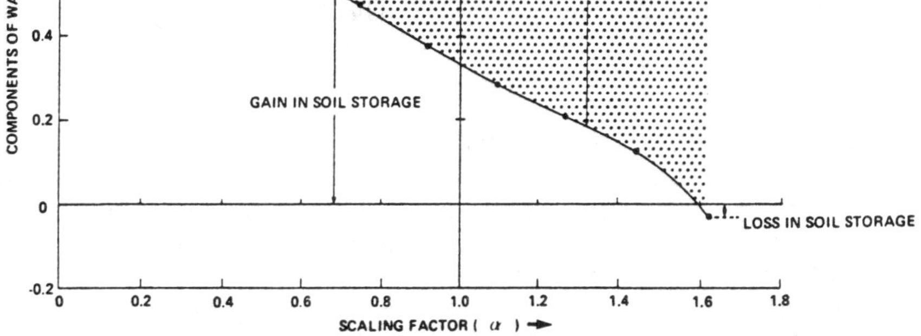

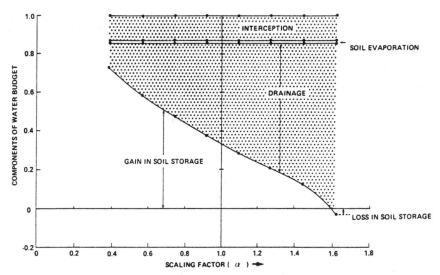

Fig. 4–8. Simulated water budgets of spatially variable Fullerton soil in April. Water storage in the 2.7-m profile at the beginning of the month is assumed to be uniform. Components of the water budget are shown as fractions of the monthly precipitation. The scaling factor α is the ratio of a microscopic length scale in the soil to its average over an area (after Peck et al., 1977).

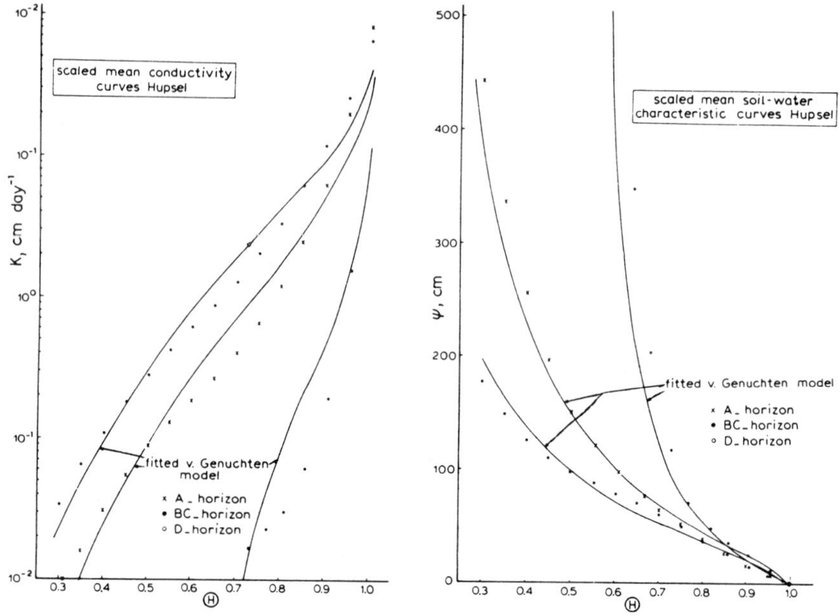

Fig. 4–9. Reference curves of A, BC horizon, and D horizon (after Hopmans and Stricker, 1989).

were to quantify how soil spatial variability impacts plant transpiration and to quantify statistical properties of the water balance components.

Soil hydraulic properties were measured at three scales of observation over the 650-ha watershed. Eventually the values were reduced to three mean reference curves for K and h as given in Fig. 4–9. The curves are defined by fitted van Genuchten parameters (Eq. [7] and [15] and profiles within the watershed described by a statistical distribution of α and θ_s.

In addition to soil hydraulic properties, the lower boundary conditions varies from spot to spot within the field. This was eventually taken as

$$q = (a_{\text{ref}}/\gamma_i) \exp (b^*|H|) \qquad [27]$$

where q is a downward flux integrated for the watershed, a_{ref} and b^* are constants, γ_i is a site specific scale factor and H the groundwater level at a specific location. For the watershed a_{ref} and b^* were determined based on two categories for each year. Their values depended on clay content within the first 1.2 m of the profile.

Simulations were performed with the model SWATRE (Belmans et al., 1981). In addition to the mentioned parameters and the lower boundary condition, daily precipitation and potential evaporation were used. These along with the thickness of the root zone were assumed uniform over the study area. The simulated root uptake is influenced by the site-specific values of h and $K(h)$.

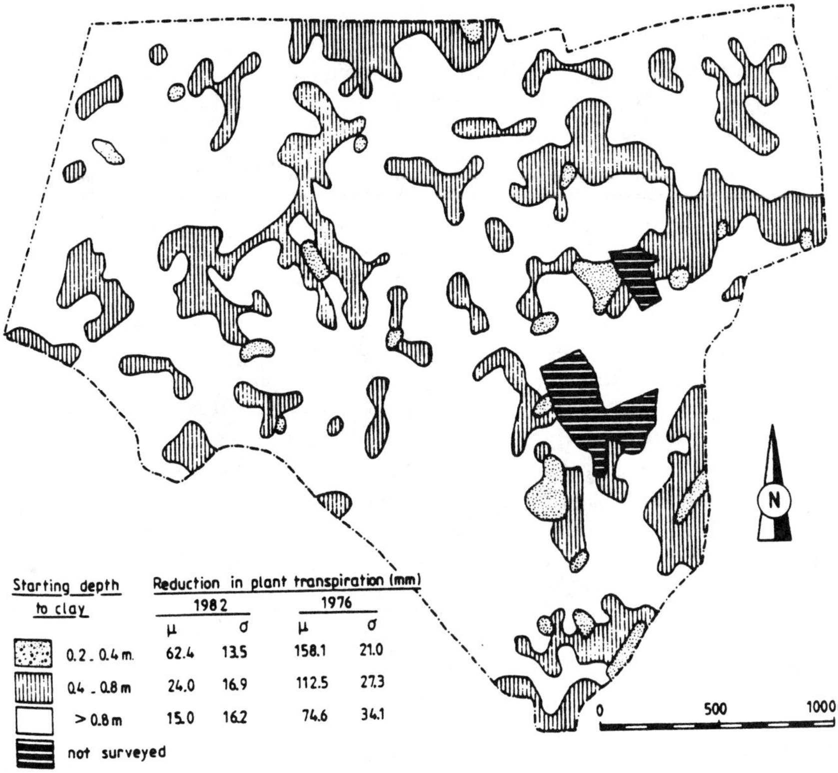

Fig. 4–10. Correspondent of ET reductions to soil classes as a result of MC simulations (after Hopmans and Stricker, 1989).

As an example of the usefulness of the results, Fig. 4–10 is a distribution of soil classes based on starting depth to clay as well as corresponding calculated reduction in plant transpiration in 1976 and 1982. The reductions are relative to potential evapotranspiration. In both years, the largest reductions occur in areas where the starting depth to clay is small and the smallest reduction occurs where the starting depth to clay is greater than 0.8 m. The trends are the same for both years even though 1976 was drier overall, resulting in much higher reductions in evapotranspiration.

DISCUSSION

Several examples have been discussed to illustrate the versatility of scaling to characterize spatial variability of soils. The first three examples are data intensive examples, ranging from a 25 m by 6 m deep trench to a 150-ha area to samples collected from many sites in the Netherlands. In each case, the conditions illustrated in Fig. 4–1 are not met, but still the heterogeneity from sample to sample was substantially captured within the scaling coefficients. In Example 4, calculations were made for general conditions and then applied to a specific case. This procedure could be used to model variability by successive choices of α. This is actually done in Example 5 (Peck et al., 1977) for calculating water balance components. In the final example, Hopmans and Stricker (1989) use the aspect of data reduction to express heterogeneity about mean values. They then follow with Monte Carlo simulations to give an output variability of water balance components as influenced by the soil variability and variation in the lower boundary condition. Finally, they relate reduction in plant transpiration to starting clay depth illustrated on a map for the 650-ha watershed.

Thus, although the idealized model of similar media has been corrupted somewhat, applications have proven quite versatile. As a closing comment, the popular contemporary topic of fractals is closely connected to scaling. The emphasis in that case is on self-similarity. This is roughly equivalent to viewing Fig. 4–1A and 1B as different scales of the same material—from closeup and from far away. Also, time-invariance as addressed by Vachaud et al. (1985) has some close analogies to scaling. For time-invariance the ranking is preserved for a property through a sequence of time. Thus, if the value of a property (such as water content) is followed through time, site values preserve their order.

REFERENCES

Belmans, C., J.G. Wesseling, and R.A. Feddes. 1981. Simulation model of a cropped soil providing different types of boundary conditions (SWATRE). Nota 1257, ICW, Wageningen.

Hills, R.G., D.B. Hudson, and P.J. Wierenga. 1989. Spatial variability at the Las Cruces Trench Site. *In* M. Th. van Genuchten (ed.) Indirect methods for estimating the hydraulic properties of unsaturated soils, proc., Riverside, CA. 11–13 October.

Hopmans, J.W., and J.N.M. Stricker. 1989. Stochastic analysis of soil water regime in a watershed. J. Hydrol. 105:57–84.

Miller, E.E. 1980. Similitude and scaling of soil-water phenomena. p. 300–318. *In* D. Hillel (ed.) *Applications of soil physics*. Academic Press, New York.

Miller, E.E., and R.D. Miller. 1955a. Theory of capillary flow: I. Practical implications. Soil Sci. Soc. Am. Proc. 19:267–271.

Miller, R.D., and E.E. Miller. 1955b. Theory of capillary flow: II. Experimental information. Soil Sci. Soc. Am. Proc. 19:271–275.

Miller, E.E., and R.D. Miller. 1956. Physical theory for capillary flow phenomena. J. Applied Phys. 27:324–332.

Moore, R.E. 1939. Water conduction from shallow water tables. Hilgardia 12:383–426.

Nielsen, D.R., J.W. Biggar, and K.T. Erh. 1973. Spatial variability of field-measured soil-water properties. Hilgardia 42:215–260.

Peck, A.J., R.J. Luxmoore, and J.L. Stolzy. 1977. Effects of spatial variability of soil hydraulic properties in water budget modeling. Water Resour. Res. 13:348–354.

Philip, J.R. 1969. Theory of infiltration. Adv. Hydrosci. 5:215–296.

Rao, P.S.C., R.E. Jessup, A.G. Hornsby, D.K. Cassel, and W.A. Pollans. 1983. Scaling soil microhydrologic properties of Lakeland and Konawa soils using similar media concepts. Agric. Water Manage. 6:277–290.

Russo, D., and E. Bresler. 1980. Scaling soil hydraulic properties of a heterogeneous field. Soil Sci. Soc. Am. J. 44:681–684.

Sposito, G., and W.A. Jury. 1985. Inspectional analysis in the theory of water flow through unsaturated soil. Soil Sci. Soc. Am. J. 49:791–798.

Vachaud, G., A.P. De Silans, P. Balabanis, and M. Vauclin. 1985. Temporal stability of spatially measured soil water probability density functions. Soil Sci. Soc. Am. J. 49:822–828.

van Genuchten, M.Th. 1980. A closed-form equation for predicting the hydraulic conductivity of unsaturated soils. Soil Sci. Soc. Am. J. 44:892–898.

Warrick, A.W., D.O. Lomen, and S.R. Yates. 1985. A generalized solution to infiltration. Soil Sci. Soc. Am. J. 49:34–38.

Warrick, A.W., G.J. Mullen, and D.R. Nielsen. 1977. Scaling field-measured soil hydraulic properties using a similar media concept. Water Resour. Res. 13:355–362.

Wösten, J.H.M. 1989. Use of scaling techniques to quantify variability in hydraulic functions of soils in the Netherlands. *In* H. Fluhler (ed.) Monte Verita Conference Proc., Monte Verità, Switzerland. 25–29 September.

5 Application of Scaling to the Analysis of Unstable Flow Phenomena

J.-Y. Parlange and T.S. Steenhuis

Cornell University
Ithaca, New York

R.J. Glass

Sandia National Laboratories
Albuquerque, New Mexico

Water infiltration into layered soils where a fine-textured soil overlays a coarse sand exhibits Taylor instability, and the flow field in the coarse sand breaks into fingers. Following the pioneering work of Saffman and Taylor (1958), Hill and Parlange (1972) studied this phenomenon experimentally and later proposed a theoretical model (Parlange & Hill, 1976). In this first experiment, heterogeneous packing of the coarse sand promoted merging of fingers resulting in a small number of near-saturated fingers. The increased water content resulting from merging of fingers is readily apparent from the experiments of Glass et al. (1989a). The theory provided a good description of finger diameters both in the laboratory and the field (Starr et al., 1978). Later, Hillel and Baker (1988) pointed out that when water first enters the coarse sand the matric potential should be expected to be negative and the finger to be unsaturated. The theory of Parlange and Hill (1976) applies equally well to that case, yielding for average finger width, D,

$$D = \left[\frac{\pi S_F^2}{K_F(\theta_F - \theta_i)}\right]\left[\frac{1}{1 - (q/K_F)}\right] \qquad [1]$$

Here, θ_i is the initial water content in the coarse sand, θ_F is the average water content in the fingers, and K_F and S_F are the corresponding conductivity and sorptivity. The experimental conditions for which Eq. [1]

holds consists of a two-dimensional chamber (where the fingers can be conveniently observed) and a constant flux q entering the chamber per unit width of the chamber. The flux is controlled by the conductivity of the fine-textured soil layer overlaying the coarse sand.

We wrote Eq. [1] as the product of two square brackets. The first one is soil-dependent and requires a knowledge of θ_F. Even with a very homogeneous soil, Glass et al. (1989a) find that mergers take place between fingers. For instance, in their Fig. 7 each finger resulted from two mergers on the average. As mentioned earlier the water content in each individual finger is primarily a function of mergers, as well as the water entry potential in the coarse sand as discussed by Hillel and Baker (1988). Figure 7 of Glass et al. (1989a) also shows that there is a slight widening of fingers with water content, i.e., with the number of mergers. This is quite consistent with the form of the first bracket in Eq. [1] and the behavior of real soils. For a real soil S_F^2 increases more rapidly than $K_F(\theta_F - \theta_i)$ as θ_F increases toward saturation. There is, however, a compensation between the two factors so that the first bracket is not too sensitive to the exact value of θ_F, as observed experimentally. This is also quite useful in the interpretation of experiments with the help of Eq. [1]; i.e., even though the number of mergers will not be identical for all fingers, it will result in only slight variations in width from finger to finger so that the average width for one experiment remains quite representative of all the fingers in that experiment.

The second bracket in Eq. [1] depends on the ratio q/K_F, i.e., depends on the boundary condition through q and the soil through K_F. As discussed by Hillel and Baker (1988) the term q/K_F is also the fraction of the soil occupied by fingers. Thus, a convenient way to check the validity of Eq. [1] is to plot D as a function of the fraction of soil wetted by fingers. Figure 5–1 shows the dependence of D on the fractional area occupied by fingers for a particular sand. Experimental points (◇) are from Glass et al. (1989b) for sand with a mean diameter of 0.0991 cm and the theory (solid line) represents Eq. [1] with the first bracket estimated for that particular sand. For this sand, $D^* = D$. Notice that according to Eq. [1], finger diameter for two sands is not equal for the same wetted area. It is clear that there is a remarkable agreement between theory and observations.

We are now in position to check the application of the Miller and Miller (1956) theory of scaling to finger instability. With this theory, the scaled, or reduced, finger diameter should be independent of the sand type.

MILLER SCALING

In a fundamental paper, Miller and Miller (1956) discussed the relationship between properties for similar soils, i.e., soils packed identically and with identical particle distributions when scaled with the average particle size.

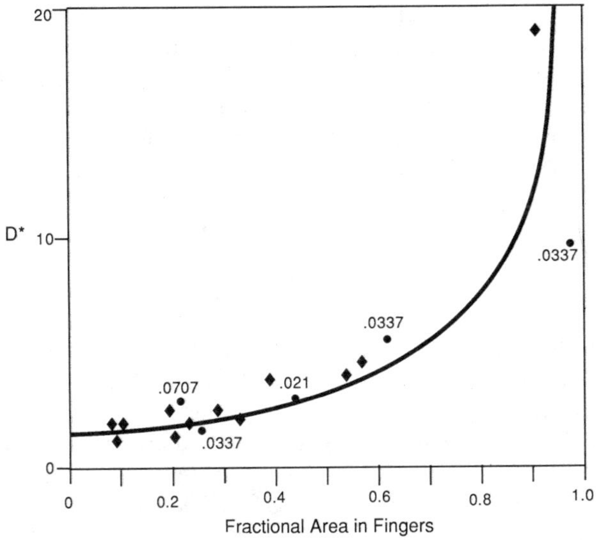

Fig. 5–1. Reduced finger width, D^*, as a function of the fractional area of wetted sand, i.e., occupied by the fingers. The diamonds correspond to $M = 0.0991$ cm, and the dots to $M = 0.0707, 0.0337$, and 0.021 cm as indicated on the figure. The solid line is the theoretical result obtained from Eq. [1] and [2].

To obtain such sands, a white silica sand used commercially for sand blasting was dry-sieved, yielding 10 sands of different mean grain sizes. Out of those, three were selected having similar particle size distributions as the sand used by Glass et al. (1989b). The average particle sizes, M, are 0.021, 0.0337, and 0.0707 cm, respectively, with the original one (Glass et al., 1989b) being $M = 0.0991$ cm. The maximum and minimum particle sizes are fairly close to the mean in each case with a distribution essentially uniform in between. Thus, the similarity of all distribution results if the spread of particle size divided by the average size is the same for all four sands. The numbers given in Table 5–1 show that this is indeed the case with a fairly good accuracy.

Under those conditions we expect the widths of the fingers to differ according to the first bracket of Eq. [1]. To calculate the dependence of this bracket on M we follow the rules for similar soils (Miller & Miller,

Table 5–1. Average particle size and relative size spread for four sands.

Mean grain diam. M	Maximum-minimum size / M
cm	
0.0991	0.330
0.0707	0.348
0.0337	0.365
0.0210	0.348

1956). The soil-water potential (pressure) Ψ, in the fingers behaves like M^{-1}, but the corresponding, θ_F, is the same. By analogy with Poiseuille flow K_F must behave like M^2, where as $S^2 \sim \int Kd\Psi$ should behave like M. Altogether then the first bracket is proportional to M^{-1}. That is, the finger width is inversely proportional to the mean grain size. To take advantage of this result, we define a reduced finger width, D^* given by

$$D^* = D0.0991/M \tag{2}$$

With this definition, for the same fractional area occupied by the fingers, all reduced finger widths, D^*, would be the same, and correspond to the physical width D when $M = 0.0991$ already plotted in Fig. 5–1.

Several experiments are reported in Fig. 5–1. It is clear that the three experiments with $M = 0.021$, 0.0337, and 0.0707 cm, respectively, when the fractional area occupied by fingers is less than 0.5 follows the reduction procedure very well. Two experiments when it is greater than 0.5 and for $M = 0.0337$ cm are also reported. The discrepancies for those two experiments are easily explained. As the fractional area increases, fingers are close to each other and some may coalesce, i.e., run parallel without actually merging. If those siamese twins are incorrectly counted as one, the result is an apparent finger width, which is greater than it should be. This is what happened when the fractional area was about 0.61. For the last experiment, q was close to the saturated conductivity so that the flow was close to being stable. In fact, the predicted finger width was several times larger than the chamber width. Under those conditions fingers could not develop. The spurious D^* indicated represents the wave length of the wavy front, which appeared in the chamber.

These last two results were given to point out that some care must be given in the interpretation of unstable flows. Luckily in practical situations, when fingers are observed in the field, the fractional area occupied by fingers is always small and those difficulties do not appear. In fact, for those field experiments the influence of the second bracket is always small.

In conclusion we have shown that Miller's scaling model can be applied with confidence to predict finger width for unstable flows. This should prove very useful in practice to easily estimate finger width in soils of a given texture and conclude whether instability could become a problem, for instance to predict the fate of contaminants.

REFERENCES

Glass, R.J., T.S. Steenhuis, and J.-Y. Parlange. 1989a. Mechanisms for finger persistence in homogeneous, unsaturated porous media: Theory and verification. Soil Sci. 148:60–70.

Glass, R.J., T.S. Steenhuis, and J.-Y. Parlange, 1989b. Wetting front instability. 2. Experimental determination of relationships between system parameters and two-dimensional unstable flow field behavior in initially dry porous media. Water Resour. Res. 25:1195–1207.

Hill, D.E., and J.-Y. Parlange. 1972. Wetting front instability in layered soils. Soil Sci. Soc. Am. Proc. 36:697–702.

Hillel, D., and R.S. Baker. 1988. A descriptive theory of fingering during infiltration into layered soils. Soil Sci. 146:51–56.

Miller, E.E., and R.D. Miller. 1956. Physical theory for capillary flow phenomena. J. Appl. Phys. 27:324–332.

Parlange, J.-Y., and D.E. Hill. 1976. Theoretical analysis of wetting front instability in soils. Soil Sci. 122:236–239.

Saffman, P.G. and G.I. Taylor. 1958. The penetration of a fluid into a porous medium or Hele-Shaw cell containing a more viscous fluid. Proc. R. Soc. London. A. 245:312–331.

Starr, J.L., H.C. de Roo, C.R. Frink and J.-Y. Parlange. 1978. Leaching characteristics of a layered field soil. Soil Sci. Soc. Am. J. 42:386–391.

6 Characteristic Lengths and Times Associated with Processes in the Root Zone

P.A.C. Raats

Institute for Soil Fertility
Haren, the Netherlands

The Millers contributed to root ecology in two areas, which are:

1. Over a period of more than 30 yr, Bob Miller studied various aspects of the mechanical interaction between roots and soils;
2. Ed Miller was involved in two key developments concerning uptake of water by plant roots from soils.

Ed Miller supervised the dissertations of Wolf (1967) on uptake of water by growing root systems, and of Herkelrath (1975) and Herkelrath et al. (1977a, b) on the influence of soil water content and soil water potential upon uptake, leading to the "root contact" model. I will return to these studies in the course of this article.

To better understand poor root growth in some compacted soils, Gill and Miller (1956) studied the roles of mechanical impedance and oxygen supply. Inspiration for this study was derived from Pfeffer's late 19th century experiments on root growth pressures exerted by plants (Gill & Bolt, 1955). To minimize the roles of other factors, Gill and Miller used glass beads and tap water as a substrate, and sterilized, freshly germinated corn (*Zea mays* L.) seedlings as the plant material. They determined root growth as a function of the applied pressure, with the percent oxygen present in the soil atmosphere as a parameter.

The mechanics of root penetration was the subject of two other studies under Bob Miller's guidance. Evans (1965) developed a model for radial growth in a granular soil. This was followed by a study of radial resistance as a function of root size and spacing by Römkens and Miller (1971).

In recent years Bob Miller had occasion to combine his intermittent interest in plant roots and his continual interest in the physics of freezing

 Scaling in Soil Physics: Principles and Applications, SSSA Special Publication no. 25.

of soils in a study of frost upheaval of overwintering plants (Perfect et al., 1987, 1988). They found that two modes of upheaval can be distinguished. On the one hand, seedlings and transplants may be uprooted by surficial soil freezing in the fall and/or spring. On the other hand, well-anchored plants are displaced by deep frost penetration during midwinter.

Following some general comments on scaling Richards' Equation, the remainder of this article will deal with various aspects of the uptake of water, including the role of poor contact between root and soil and uptake by a growing root system.

SOME ASPECTS OF SCALING

To describe movement of water in unsaturated soils, nearly 60 yr ago Richards (1931) proposed the simplest possible balance of mass and balance of momentum, the latter expressed in terms of Darcy's Law. The balance of mass for the water may be written as

$$\delta\theta / \delta t = \underset{\sim}{\nabla}(\theta \underset{\sim}{v}) - u \qquad [1]$$

where t is the time, $\underset{\sim}{\nabla}$ is the vector differential operator, θ is the volumetric water content, $\underset{\sim}{v}$ is the velocity of the water, and u is the volumetric rate of uptake. The volumetric flux $\theta\underset{\sim}{v}$ is given by Darcy's Law:

$$\theta \underset{\sim}{v} = -k[h]\underset{\sim}{\nabla} h + k[h]\underset{\sim}{\nabla} z \qquad [2a]$$

$$= -D[\theta]\underset{\sim}{\nabla}\theta + k[\theta]\underset{\sim}{\nabla} z \qquad [2b]$$

$$= -\underset{\sim}{\nabla}\phi + k[\phi]\underset{\sim}{\nabla} z \qquad [2c]$$

where h is the tensiometer pressure head, z is a vertical coordinate with its origin at the soil surface and taken position downward, and the diffusivity D and the matric flux potential ϕ are defined by

$$D = k dh / d\theta \qquad [3]$$

$$\phi - \phi_0 = \int_{h_0}^{h} k dh = \int_{\theta_0}^{\theta} D d\theta \qquad [4]$$

Symbols in brackets denote functional dependence. Unlike the dependence of k upon θ, the dependence of h upon θ is subject to hysteresis. As a consequence, Eq. [2b] and [2c] are, strictly, only valid for monotonic changes in water content from some initial condition with uniform θ and h.

The retention and conduction of water by soils are primarily governed by the relationships between h and θ, and between k and θ. These relationships vary widely among soils. The scaling theory of Miller and Miller (1956) is concerned with geometrically similar media characterized by length scales $\lambda_* = 1$ and λ. Figure 6–1 shows two geometrically similar media with geometrically similar distributions of water and of air. For such a pair, the STVF (surface tension, viscous flow)–theory implies that simple relationships exist between the pairs of water contents, pressure heads, and hydraulic conductivities:

1. Geometric similarity implies

$$\theta = \theta_* \qquad [5]$$

2. The inverse relationship between the pressure head and the mean radius of curvature implies

$$h = \lambda^{-1} h_* \qquad [6]$$

3. The linearized Navier-Stokes Equation at the microscopic scale implies that in Darcy's Law at the macroscopic scale the hydraulic conductivity satisfies

$$k = \lambda^2 k_* \qquad [7]$$

The three scaling rules just given are of the form (Raats, 1983)

$$v = \lambda^n v_* \qquad [8]$$

with integer n. Using the three primary scaling rules, secondary scaling rules can be inferred from Darcy's Law and from the volume balance for the water. Darcy's Law implies simple scaling rules for the spatial coordinates x, y, and z, the velocity $\underset{\sim}{v}$, the volumetric flux $\theta \underset{\sim}{v}$, the total head $H = h + z$, the diffusivity D, and the matric flux potential ϕ. The volume balance for the water implies scaling rules for the time t, and the volumetric

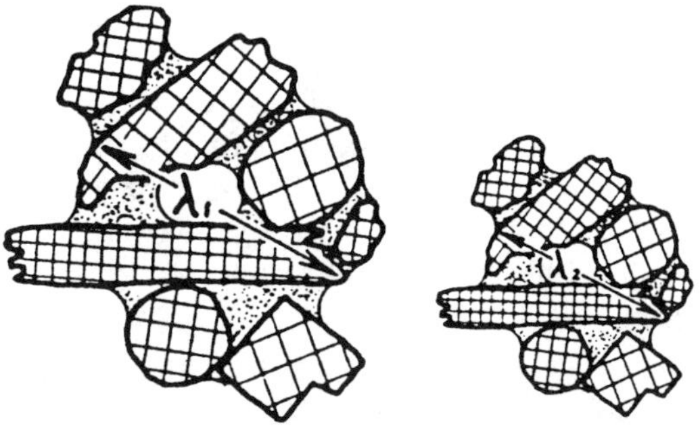

Fig. 6–1. Two geometrically similar media in similar states (Miller & Miller, 1956).

Table 6–1. Scaling rules and implied means and variances for a set of similar media with lognormally distributed length scales.

v	n	Mean of v	Variance of v
t	−3	-3μ	$9\sigma^2$
h	−1	$-\mu$	σ^2
H	−1	$-\mu$	σ^2
x, y, z	−1	$-\mu$	σ^2
θ	0	—	—
∇	1	μ	σ^2
$d\theta/dh$	1	μ	σ^2
D	1	μ	σ^2
ϕ	1	μ	σ^2
$k^{-1}\,dk/dh$ (= a for A in Table 6–2)	1	μ	σ^2
k	2	2μ	$4\sigma^2$
θv	2	2μ	$4\sigma^2$
v	2	2μ	$4\sigma^2$
$s = dk/d\theta$	2	2μ	$4\sigma^2$
dk/dh	3	3μ	$9\sigma^2$
u	3	3μ	$9\sigma^2$

rate of uptake u. Scaling rules for the water capacity $d\theta/dh$, the characteristic inverse length $a = k^{-1}dk/dh$, and the characteristic speed $s = dk/d\theta$, all three potentially a function of the water content, can also be inferred easily. Table 6–1 gives the values of n in Eq. [8] associated with the various parameters. Most noteworthy are the scaling rules for the spatial coordinates and the time.

1. The spatial coordinates and hence all macroscopic length scales in processes, should be inversely proportional to the microscopic length scale.
2. The time coordinate, and hence all time scales in processes, should be inversely proportional to the cube of the microscopic length scale.

An important implication of the power function dependence on λ^n of all the variables v in Table 6–1 is that if the length scale λ is lognormally distributed, then all the variables v/v_* will also be lognormally distributed (Raats, 1983). This is a consequence of the reproductive rule for lognormal distributions: if the variable χ is lognormally distributed with mean μ and variance σ then $e^a\chi^b$ is lognormally distributed with mean $a + b\mu$ and variance $(b\sigma)^2$.

The STVF theory of Miller and Miller concerns classes of geometrically similar media. An alternative method of defining classes of similar media is to describe the relationships among the water content θ, the pressure head h, and the hydraulic conductivity k for such classes. In effect this is often done in terms of parametric expressions for these relationships. Important examples are (see Table 6–2):

1. The Class of Mildly Nonlinear Soils with Linear $h[\theta]$ and Exponential $k[\theta]$ Relationships, Implying Exponential $D[\theta]$, $k[h]$, and $D[h]$ Relationships (Raats, 1983). The exponential $k[h]$ relationship linearizes

Table 6–2. Two classes of soils.

		Mildly nonlinear soils (A)	Power function soils (B)
Primary relationships	$h[\theta]$	$h_r + \gamma(\theta - \theta_r)$	$h_a(\theta/\theta_s)l$
	$k[\theta]$	$k_r \exp \beta(\theta - \theta_r)$	$k_a(\theta/\theta_s)^m$
Derived relationships	$D[\theta]$	$D_r \exp \beta(\theta - \theta_r)$ where $D_r = \gamma k_r$	$D_a(\theta/\theta_s)^n$ where $D_a = (lk_s h_a/\theta_s)$, $n = l + m - 1$
	$k[h]$	$k_r \exp a(h - h_r)$ where $a = \beta/\gamma$	$k_s(h/h_a)^p$ where $p = m/l$
	$D[h]$	$D_r \exp a(h - h_r)$	$D_s(h/h_a)^q$ where $q = n/l = (l + m - 1)/l$

the gravitational term in Darcy's Law expressed in terms of the matric flux potential and as such has been extremely useful in obtaining analytical solutions of steady flow problems, including problems involving uptake of water by plant roots (e.g., Raats, 1974a, 1976). Rereading Miller and Miller (1956), I noticed that they already pointed out that "$C(p)\{= a\}$ alone fully describes steady-flow behavior" and that "it may even be possible to approximate $C(p)\{= a\}$ by a constant for some purposes." For mildly nonlinear soils, the parameter a is proportional to the length scale λ.

2. The Class of Power Function Soils. As will be exemplified in this chapter, this class can in some cases also be used to obtain analytical solutions. The class of power function soils can be seen as a subclass of a superclass of soils, which shares flexibility with a rather sound basis in Poiseuillian flow in networks of capillaries (Raats, 1990). Members of this superclass are regularly used in numerical studies and as a basis for interpreting laboratory and field observations. For power function soils, the air entry pressure head h_a is inversely proportional to the length scale λ.

In the abstract for his lecture at Las Vegas, Ed Miller encourages "the use of the microlength λ as a natural part of any parameterized description of soil properties" (Miller, 1989). It may well be that this idea originated on a Northwest Orient flight Ed and I took sometime in 1969 (See Fig. 6–2).

SCALING OF UPTAKE

We have already seen that introducing the three basic scaling rules in the balance of mass shows that similarity requires that the volumetric rate of uptake is taken proportional to the cube of the length scale λ. This requirement is satisfied if the rooting depth is taken inversely proportional to λ and if the rate of transpiration is taken proportional to λ^2. The role of the rooting depth can be nicely demonstrated by considering the volumetric rate of uptake to be given by (Raats, 1974a, 1976)

$$u = f[z]T \tag{9}$$

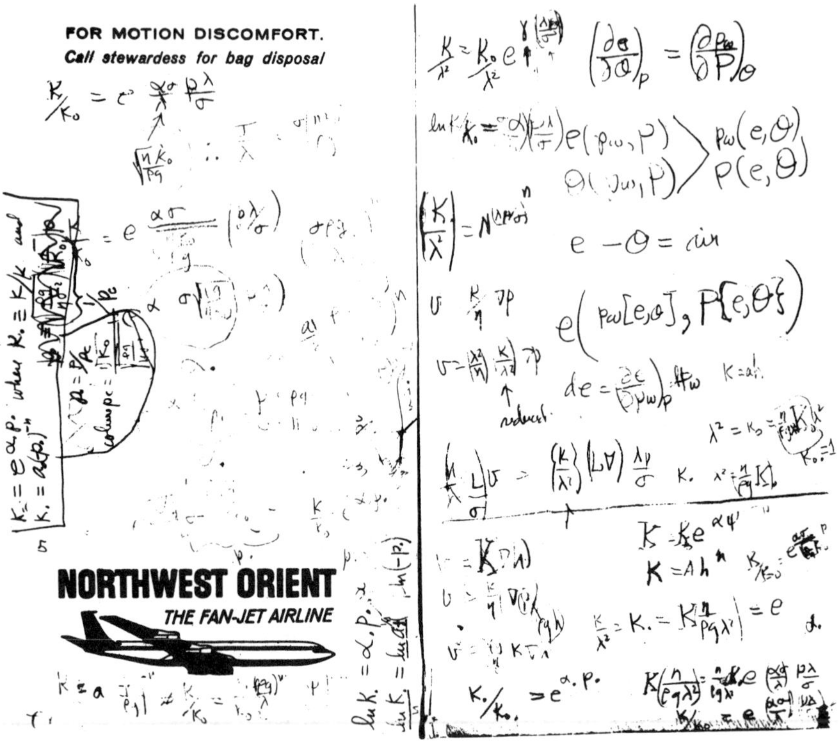

Fig. 6–2. Notes by Ed Miller, made sometime in 1969, on scaling of the exponential and power function dependencies of the hydraulic conductivity upon the soil water pressure. Note that, just as in Miller and Miller (1956), a reduced variable is denoted by a period following a symbol. Note further that, during the same flight, the equation of state for water in nonrigid soils was discussed briefly.

where T is the rate of transpiration and $f[z]$ is the distribution function for the uptake of water by plant roots. The distribution function

$$f = \delta^{-1} \exp(-z/\delta) \qquad [10]$$

where δ can be interpreted as an equivalent rooting depth, has turned out to be rather useful (Raats, 1974a, 1976). Together with the exponential $k[h]$ relationship for mildly nonlinear soils, this distribution function can be used to obtain solutions for steady flow involving uptake by plant roots. Specifically, for a constant rate of infiltration F_0 and in the absence of a water table, the distribution of the matric flux potential is given by (Raats, 1974a)

$$a\phi/F_0 = L + (1 - L)\frac{a\delta}{1 + a\delta}\exp(-z/\delta) \qquad [11]$$

where $L = (F_0 - T)/F_0$ is the leaching fraction. The dimensionless number

$a\delta$ embodies the interaction of the length scales of the soil and the root system. Taking a as the length scale of the soil, Eq. [11] can be written as

$$\phi_*/F_{0*} = L + (1 - L)\frac{\delta_*}{1 + \delta_*}\exp(-z_*/\delta_*) \qquad [12]$$

In the example just given it is assumed that the plant has a certain demand for water and that this demand can be met at all times. This is certainly not always the case. In water balance models such as SWATRE, limited availability of water when the soil is either too wet or too dry is taken into account (cf., Feddes et al., 1978; Belmans et al., 1983).

Individual roots function at a meso-scale, which is intermediate between the microscopic Navier-Stokes scale and the macroscopic scale at which the uptake by plant roots is averaged over a large number of roots, as in Eq. [1]. Ever since the pioneering studies of Philip (1957) and Gardner (1960), mass balance Eq. [1] with u omitted and Darcy Eq. [2] have been used to analyze the movement of water in regions affected by individual roots. At this meso-scale of individual roots, even the simplest model of uniformly distributed parallel roots requires two length scales, r_0, the radius of the root, and r_1, the outer radius of the hollow cylinder of soil associated with the root. With flow to individual roots are also associated two characteristic times, t_d and $t_{s/d}$ defined by

$$t_d = r_1^2/\overline{D}, \qquad t_{s/d} = (1 - \rho_0^2)\theta_i B/T \qquad [13]$$

where $\overline{D}$ is the mean of the soil water diffusivity in the appropriate range, θ_i is the initial water content, and B is the rooting depth. The time t_d characterizes the diffusive transport of the water to the root. The time $t_{s/d}$ arises from the ratio of the supply $(1 - \rho_0^2)\theta_i$ of water in the soil and the demand T/B by the plant, where $\rho_0 = r_0/r_1$.

To describe the flow to an individual root, it is convenient to introduce the dimensionless radial coordinate ρ, time τ, soil water depletion Δ, and diffusivity $\mathfrak{D}$:

$$\rho = r/r_1 \qquad \tau = t/t_d \qquad [14]$$

$$\Delta = (\theta_i - \theta)/\theta_i \qquad \mathfrak{D} = D/\overline{D} \qquad [15]$$

It turns out that the length scale r_0 and the time scale $t_{d/s}$ occur in the flow problem through the dimensionless parameters ρ_0 and $\tau_{s/d}$ defined by

$$\rho_0 = r_0/r_1 \qquad \tau_{s/d} = t_{s/d}/t_d \qquad [16]$$

In terms of the dimensionless variables the uptake problem can be stated as in Table 6–3. Up to the dimensionless time τ_{crit}, the solution of the flow problem depends on the soil property $\mathfrak{D}[\Delta]$ and the parameters ρ_0

Table 6–3. Uptake of water by a plant root in terms of dimensionless variables.

$\frac{\delta \Delta}{\delta \tau} = \frac{\delta}{\delta \rho} \mathcal{D} \frac{\delta \Delta}{\delta \rho}$		
$\tau = 0$	$\rho_0 < \rho < 1$	$\Delta = 0$
$\tau > 0$	$\rho = 1$	$\frac{\delta \Delta}{\delta \rho} = 0$
A. Constant rate of uptake		
$0 < \tau < \tau_{\text{crit}}$	$\rho = \rho_0$	$\mathcal{D}\frac{\delta \Delta}{\delta \rho} = \frac{1 - \rho_0^2}{2\rho_0} \frac{t_d}{t_{d/s}}$
B. Falling rate of uptake		
$\tau > \tau_{\text{crit}}$	$\rho = \rho_0$	$\Delta \rightarrow \Delta_{\text{lim}}$

and $\tau_{s/d}$. In particular this means that τ_{crit} will depend on $\mathcal{D}[\Delta]$, ρ_0, and $\tau_{s/d}$. The evaluation of τ_{crit} is the central point of interest in the analysis of uptake by plant roots in the Ph.D. thesis of de Willigen and van Noordwijk (1987). For $\tau > \tau_{\text{crit}}$ the solution may eventually be governed mainly by the ability of the soil to supply water to the soil/root interface.

Stating the flow problem in terms of dimensionless variables greatly increases the efficiency of computations, because the six variables θ_i, D_1, r_0, r_1, T, and B have been coalesced into the three variables $\mathcal{D}_0$, ρ_0, and $\tau_{s/d}$. Further efficiency is obtained by considering Miller scaling. Any solution of the uptake problem stated in Table 6–3 applies to any Miller similar flow. Similarity requires that the length scales r_0, r_1, and B are proportional to λ^{-1}, that the rate of transpiration is proportional to λ^2, and that the time scales t_d and $t_{s/d}$ are proportional to λ^{-3}. Therefore, the coarser the soil, the thinner and more closely spaced the individual roots should be, the smaller the rooting depth should be, and the more rapid the flow process should evolve. Horticulturalists create a wide range of root environments, which tend to being Miller similar: they use coarse substrates, in thin layers, inhabited by dense root systems, being irrigated frequently.

LIMITED CONTACT BETWEEN ROOT AND SOIL

Figure 6–3 shows schematically the poor contact between root and soil. Herkelrath et al. (1977b) suggested that as long as the potential transpiration can be met, the transport from the soil to the xylem is described by

$$0 < t < t_{\text{crit}}, R = R_0, D \frac{\delta \theta}{\delta r} = \frac{r_1^2}{2 r_0 \delta} E_{\text{pot}} = C \frac{\theta_0}{\theta_s} \{h[\theta_0] - h_{\text{xylem}}\} \quad [17]$$

where C is the conductance of the region between the soil and the xylem. The degree of saturation θ_0/θ_s of the soil at the soil/root interface is a

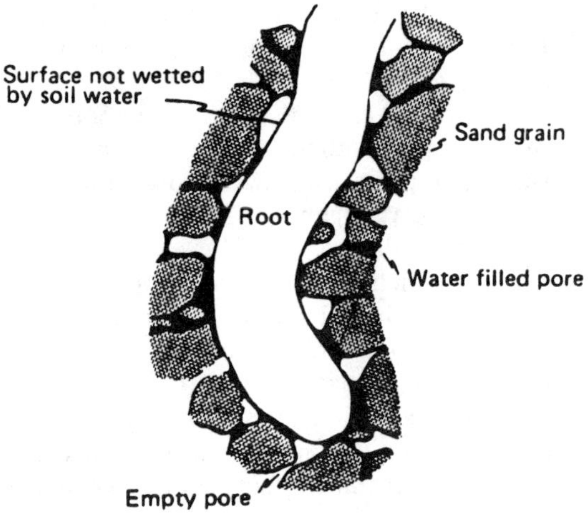

Fig. 6–3. Limited contact between root and soil (Herkelrath et al., 1977b).

factor accounting for the poor contact. From Eq. [17] it follows that at time t_{crit}

$$h[\theta_{0\mathrm{crit}}] = h_{\mathrm{xylem\ crit}} + \frac{r_1^2}{2r_0\delta}\frac{\theta_s}{\theta_{0\mathrm{crit}}}\frac{E_{\mathrm{pot}}}{C} \qquad [18]$$

or in scaled form

$$h_*[\theta_{*0\mathrm{crit}}] = h_{*\mathrm{xylem\ crit}} + \frac{r_{*1}^2}{2r_{*0}\delta_*}\frac{\theta_{*s}}{\theta_{*0\mathrm{crit}}}\frac{E_{*\mathrm{pot}}}{C_*} \qquad [19]$$

where

$$C = \lambda^3 C_* \qquad [20]$$

This means that Miller scaling requires that the coarser the soil, roots should become not only thinner and more closely spaced, but their cortex should become more permeable.

For given values of $h_{*\mathrm{xylem\ crit}}$, r_{*1}, r_{*0}, θ_{*s}, $E_{*\mathrm{pot}}$, and C_*, Eq. [19] relates the pressure head $h_{*\mathrm{crit}}$ to the water content $\theta_{*0\mathrm{crit}}$ of the soil at the soil/root interface. The infinity of pairs ($h_{0\mathrm{crit}}$, $\theta_{0\mathrm{crit}}$) or ($h_{*0\mathrm{crit}}$, $\theta_{*0\mathrm{crit}}$) is reduced to the single pair by determining the intersection of Eq. [18] or [19] with the soil water retention curve. For ($h_{0\mathrm{crit}}$, $\theta_{0\mathrm{crit}}$) this was done graphically by van Noordwijk (1983; see also de Willigen & van Noordwijk, 1987). A related graphical technique with the role of the retention curve replaced by the relationship between the hydraulic conductivity and the pressure head was used earlier in an analysis of steady infiltration into crusted soils (Raats, 1974b). Except for the presence of θ_0/θ_s, the "root contact model" is analoguous to the model commonly used to describe flow across a crust.

An alternative approach to modeling limited soil/root contact is to consider flow in a plane perpendicular to a root, assuming that the circumference consists of two parts, one part in contact with wet soil, another part in contact with air (de Willigen & van Noordwijk, 1987). The availability of the water is then reduced by the change from a purely radial flow pattern to a pattern in which also angular components of the flux are involved. The smaller the root/soil contact, the smaller the fraction of the potentially available water that can be acquired at a certain rate.

Some consequences of partial soil/root contact have been analyzed by de Willigen and van Noordwijk (1987) not only for uptake of water, but also for uptake of plant nutrients and for exchange of gases with the soil atmosphere. Although partial contact reduces the availability of water and nutrients, it enhances the exchange of gases between roots and soil atmosphere. In structured soils, roots have a tendency to follow macropores in the form of cracks, worm holes, and holes left behind by decayed roots. To what a degree partial contact is a consequence of the inability of roots to penetrate or an innate strategy assuring proper future functioning may be difficult to determine. Clustering is another feature of root distributions, especially in structured soils, limiting availability of water and nutrients (de Willigen & van Noordwijk, 1987).

UPTAKE OF WATER BY A GROWING ROOT SYSTEM

Wolf (1967) analyzed the uptake of water at a root front in an infinite, uniform soil. He discussed three cases: (i) transient flow to a stationary root front, (ii) steady flow to a moving root front, and (iii) transient flow to a moving root front. In the following, some aspects of this problem will be discussed.

Assume that the entire water uptake occurs at a plane densely populated with root tips, moving at a velocity v_f, and that the water moves in the direction z perpendicular to this plane. To discuss this class of flows, it is convenient to introduce a coordinate frame of reference that moves with the root front. Equation [21] defines the moving coordinate Z in terms of the stationary coordinate z, the time t, and the velocity of the root front v_f.

$$Z = z - v_f t \qquad [21]$$

The corresponding transformations of the space and time derivatives are

$$\frac{\delta\cdot}{\delta Z} = \frac{\delta\cdot}{\delta z}, \qquad \frac{\delta\cdot}{\delta t|_Z} = \frac{\delta\cdot}{\delta t|_z} + v_f \frac{\delta\cdot}{\delta z} \qquad [22]$$

In terms of Z and t, Eq. [1] becomes

$$\frac{\delta\theta}{\delta t|_Z} = \frac{\delta}{\delta Z}\,\theta(v - v_f) \qquad [23]$$

Assuming that behind the root front the velocity of the water is zero, the mass balance at the root front reduces to (cf., Raats, 1972)

$$\theta^+(v^+ - v_f) = -\hat{u} - \theta^- v_f \qquad [24]$$

On the left-hand side of Eq. [24] appears the flux of water relative to the root front. On the right-hand side of Eq. [24], the first term represents the rate of uptake of water per unit area, and the second term represents the flux of water into the zone behind the root front.

In terms of Z and t, Eq. [2] becomes

$$\theta v = -D\,\frac{\delta\theta}{\delta Z} \qquad [25]$$

where the gravitational force acting on the water has been neglected. At the root front the pressure head and therefore also the water content will be continuous

$$\theta^+ = \theta^- = \theta_0 \qquad [26]$$

Introducing Eq. [26] in Eq. [24] gives

$$\theta^+ v^+ = -\hat{u} \qquad [27]$$

Equations [23] and [25], together with the initial condition $\theta[Z, t] = \theta_i$ and the boundary conditions, Eq. [26] and [27], describe the flow to a root front.

Transient flow to a stationary root front can be treated by means of the so-called Boltzmann transformation. The cumulative uptake increases as $t^{1/2}$, the rate of uptake decreases as $t^{-1/2}$. The details of the pressure head and water content distributions at successive times depend on the physical properties of the soil.

With a stationary, plane root front the flow does not tend to become steady. If the root front is moving, however, the flow does tend to become steady in the frame of reference moving with the root front. When this happens, Eq. [23] reduces to

$$\frac{\delta}{\delta Z}\,\theta(v - v_f) = 0 \qquad [28]$$

Integration of Eq. [28] gives

$$\theta\,(v - v_f) = c \qquad [29]$$

Because for $Z \rightarrow \infty$, $\theta \rightarrow \theta_i$ and $v \rightarrow 0$

$$c = -\theta_i v_f \qquad [30]$$

From Eq. [25], [29], and [30] it follows that

$$-D\frac{\delta\theta}{\delta Z} = -(\theta_i - \theta)v_f \qquad [31]$$

On the left-hand side of Eq. [31] appears the flux at any Z. This flux ranges from its maximum value $-(\theta_i - \theta_0)v_f$ at $Z = 0$ to zero at $Z \rightarrow \infty$.

Integration of Eq. [31] gives

$$\frac{Zv_f}{D_0} = -\int_{\ln(\theta_i-\theta_0)}^{\ln(\theta_i-\theta)} (D/D_0)d\ln(\theta_i - \theta) \qquad [32]$$

To integrate Eq. [32], D must be known as a function of ln $(\theta_i - \theta)$. Following are closed solutions for three $D[\theta]$ functions

1. Linear soil with $D = D_0$

$$\frac{Zv_f}{D_0} = -\ln\frac{\theta_i - \theta}{\theta_i - \theta_0} \qquad [33]$$

According to Eq. [33] the dimensionless water content $(\theta_i - \theta)/(\theta_i - \theta_0)$ is an exponential function of the dimensionless distance Zv_f/D_0. This solution was also given by Wolf (1967).

2. Mildly nonlinear soils (see Table 6–2)

$$\frac{Zv_f}{D_0} = \exp\beta(\theta_i - \theta_0)\{(-E_i[-\beta(\theta_i - \theta)]) - (-E_i[-\beta(\theta_i - \theta_0)])\} \qquad [34]$$

This solution is new.

3. Power function soils (see Table 6–2)

$$\frac{Zv_f}{D_0} = -\left(\ln\frac{\theta_i - \theta}{\theta_i - \theta_0} + \sum_{n=1}^{p}\frac{1}{n}\frac{\theta^n - \theta_0^n}{\theta_i^n}\right) \qquad [35]$$

For $p = 0$, equation reduces to Eq. [33]. For $p = 1$, 2, and 4, Eq. [35] reduces to equations given by Wolf (1967). Figure 6–4 shows observed and calculated distributions of water content behind and ahead of a root front.

A quantity of particular interest is the depletion W ahead of moving root front

$$W = \int_0^\infty(\theta_i - \theta)dZ = \int_0^\infty\frac{D}{v_f}\frac{d\theta}{dZ}dZ$$

$$= \int_{\theta_0}^{\theta_i} Dd\theta/v_f = \int_{h_0}^{h_i} kdh/v_f = \frac{\phi_i - \phi_0}{v_f} \qquad [36]$$

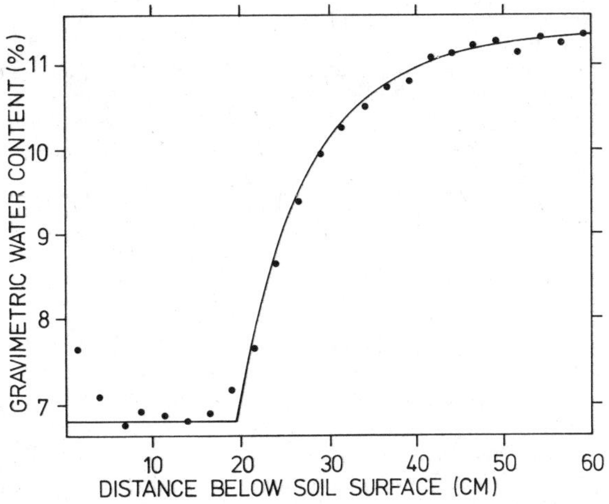

Fig. 6–4. Observed (data points) and calculated (curve) distribution of water content behind and ahead of a root front located at 20 cm below the soil surface (adapted from Wolf, 1967).

where Eq. [31] has been used. The simple evaluations of W for linear, mildly nonlinear, and power function soils are left to the reader. Equation [36] states that the effectiveness of the matric flux potential difference $\phi_i - \phi_0$ for delivering water to the root front from the region not yet explored by the root system is inversely proportional to the velocity v_f of the root front.

CONCLUDING REMARKS

Thirty years ago, when I was a M.Sc. student at Wageningen Agricultural University, Gerry Bolt asked me to determine and explain water retention curves of mixtures of sand and montmorillonite. This called for study of two, then recent developments in soil science: double layer theory for clays and STVF-theory for sands (and glass beads). I am grateful that at such an early stage I was introduced to those two far-reaching physical-mathematical models and learned about their limitations. Despite the fact that I hardly added to what Bolt and Miller (1958) had already written, theories of such calibre became a lasting interest.

REFERENCES

Belmans, C., J.G. Wesseling, and R.A. Feddes. 1983. Simulation model of the water balance of a cropped field: SWATRE. J. Hydrol. 63:271–286.

Bolt, G.H., and R.D. Miller. 1958. Calculation of total and component potentials of water in soil. Trans. Am. Geophys. Union 39:917–928.

de Willigen, P., and M. van Noordwijk. 1987. Roots, plant production and nutrient use efficiency. Ph.D. thesis. Agricultural Univ., Wageningen, the Netherlands.

Evans, G.N. 1965. A model of radial root growth in a granular soil Ph.D. thesis. Cornell Univ. (Diss. Abstr. 26:2943).

Feddes, R.A., P.J. Kowalik, and H. Zaradny. 1978. Simulation of field water use and crop yield. PUDOC, Wageningen, the Netherlands.

Gardner, W.R. 1960. Dynamic aspects of water availability to plants. Soil Sci. 89:63–73.

Gill, W.R., and G.H. Bolt. 1955. Pfeffer's studies of the root growth pressures exerted by plants. Agron. J. 47:166–168.

Gill, W.R., and R.D. Miller. 1956. A method for study of the influence of mechanical impedance and aeration on the growth of seedling roots. Soil Sci. Soc. Am. Proc. 20:154–157.

Herkelrath, W.N. 1975. Water uptake by plant roots. Ph.D. thesis. Univ. of Wisconsin (Diss. Abstr. 36:3710).

Herkelrath, W.N., E.E. Miller, and W.R. Gardner. 1977a. Water uptake by plants: I. Divided root experiments. Soil Sci. Soc. Am. J. 41:1033–1038.

Herkelrath, W.N., E.E. Miller, and W.R. Gardner. 1977b. Water uptake by plants. II. The root contact model. Soil Sci. Soc. Am. Proc. 41:1039–1043.

Miller, E.E. 1989. Perspectives on scaling in soil physics. p. 189–190. *In* Agronomy abstracts. ASA, CSSA, and SSSA, Madison, WI.

Miller, E.E., and R.D. Miller. 1956. Physical theory for capillary flow phenomena. J. Appl. Phys. 27:324–332.

Perfect, E., R.D. Miller, and B. Burton. 1987. Root morphology and vigor effects on winter heaving of established alfalfa. Agron. J. 79:1061–1067.

Perfect, E., R.D. Miller, and B. Burton. 1988. Frost upheaval of overwintering plants: A quantitative field study of the displacement process. Arct. Alp. Res. 20:70–75.

Philip, J.R. 1957. The physical principles of soil water movement during the irrigation cycle. p. 8.125–8.154. *In* Commission on Irrigation and Drainage Third Congress. Int. Comm. on Irrig. and Drainage, San Francisco, CA.

Raats, P.A.C. 1972. Jump conditions in the hydrodynamics of porous media. *In* Fundamentals of Transport Phenomena in Porous Media, IAHR-ISSS Symp. at Guelph, Ontario 1:155–173. 7–11 August. Univ. of Guelph, Guelph, ON, Canada.

Raats, P.A.C. 1974a. Steady flows of water and salts in uniform soil profiles with plant roots. Soil Sci. Soc. Am. Proc. 38:717–722.

Raats, P.A.C. 1974b. Steady infiltration into crusted soils. 10th Int. Congress of Soil Sci. (Moscow) Trans. 1:75–80.

Raats, P.A.C. 1976. Analytical solutions of a simplified flow equation. Trans. ASAE 19:683–689.

Raats, P.A.C. 1983. Implications of some analytical solutions for drainage of soil water. Agric. Water Manage. 6:161–175.

Raats, P.A.C. 1990. A superclass of soils. Proceedings of the International workshop on "Indirect methods for estimating the hydraulic properties of unsaturated soils," Riverside, CA. 11–13 Oct. 1989. USDA-ARS, U.S. Salinity Lab., Riverside, CA.

Richards, L.A. 1931. Capillary conduction of liquids through porous mediums. Physics 1:318–333.

Römkens, M.J.M., and R.D. Miller. 1971. Predicting root size and frequency from one-dimensional consolidation data—A mathematical model. Plant Soil 35:237–248.

van Noordwijk, M. 1983. Functional interpretation of root densities in the field for nutrient and water uptake. p. 207–226. *In* Wurzelökologie und ihre Nutzanwendung/Root Ecology and its Practical Application. Int. Symp. Gumpenstein, 1982, Bundesanstalt Gumpenstein, A-8952 Irdning. Published by Bundesanstallt für alpenländische Landwirtschaft, A–8952 Irdning.

Wolf, J. 1967. The role of root growth in supplying moisture to plants. Ph.D. thesis. The Univ. of Rochester, New York (Diss. Abstr. 68–15869).

7 Scaling of Mechanical Forces and Stresses in Unsaturated Granular Soils

Victor A. Snyder
Agricultural Experiment Station
University of Puerto Rico
Mayagüez, Puerto Rico

The resistance of unsaturated soils to deformation under applied mechanical stresses remains one of the most poorly understood aspects of soil physics. The problem is significant, because many important soil management decisions require knowledge of soil reaction to mechanical loads. Particularly important is understanding the interactions between the soil solid and pore phases as they affect soil behavior.

The theoretical description of the internal force systems and constitutive (stress-deformation) properties of mechanically loaded unsaturated soils, which determine their deformation or fracture under loading, is very complex. Internal forces are generated not only by applied external loads, but also by stresses in the pore-air and pore-water phases and in the air-water interface. The way in which these component stresses interact to cause particle movement (soil deformation) depends on the respective magnitudes of the stresses, and also on the geometrical configuration (structure) of the particle framework and pore phases. Structural geometry determines the directions and points of action of the individual force components. It also establishes the constraints of the loaded soil system; i.e., it determines the restrictions on movement of the individual particles or groups of particles within the system. For example, a soil particle that is in close contact with many neighboring particles will have less degrees of freedom (less possibilities for movement) than a particle in contact with only a few particles. The magnitudes and directions of the external and internal forces capable of resulting in particle movement will be very different in each case. Thus, structural geometry plays a major role in determining soil constitutive properties. One of the challenges still facing soil mechanics today is the development of a theory that can quantitatively describe soil structure interms of its effects on force interactions and constitutive properties of unsaturated soils.

SIMILITUDE CONSIDERATIONS

An important consideration in any physical problem is analysis of its scaling properties. In the field of soil physics, this has perhaps been best exemplified by the Miller and Miller (1956) scaling theory of capillary flow phenomena. By appropriate scaling of the Navier-Stokes equations and the Laplace surface tension equation, Miller and Miller arrived at general conclusions on soil water behavior as a function of system dimensions, the material properties of the pore-water and the body forces acting on the system. Their theory yielded much insight into the hydraulic behavior of soils with different pore sizes and, with some modification, has also proved useful in explaining the variability of soil hydraulic properties in the field, even in soils with considerable colloidal activity where capillary scaling theory in principle does not apply (Reichart et al., 1972; Warrick et al., 1977).

A considerable amount of scaling research has also been conducted on the mechanical behavior of soil-machine systems (Freitag et al., 1977). Most of this work is based on dimensional analysis, which essentially involves listing the variables that enter a problem and trying to group them into dimensionless parameters. An advantage of the technique is that knowledge of the equations that describe a system is not always necessary, provided one lists all the variables involved and complex interactions between variables with the same dimensions do not enter the problem. The procedure has strong limitations, however, if these conditions are not met. An unresolved problem in soil-machine studies, which has limited the application of similitude analysis, is precisely a lack of understanding of the appropriate soil variables that need to be included in the system (Freitag et al., 1977).

A possibility that does not appear to have been fully explored in this respect is the application of the Miller and Miller scaling theory to the problem of soil mechanical strength. It has long been observed that microscopic scale (particle and pore size) is very influential on the changes of soil strength with moisture content. Soils with small particles are generally observed to get "harder" on drying than soils with large particles. The importance of particle size to soil mechanical behavior is also highlighted by the fact that many early estimates of soil texture were based on observed mechanical behavior rather than on direct measurement of particle size (Towner, 1972).

Pioneering work on the scaling nature of soil strength was conducted many years ago by Haines (1925) and Fisher (1926), who were investigating the physical mechanisms behind the phenomenon of soil hardening upon drying. Haines and Fisher reasoned that cohesive stresses in unsaturated capillary (noncolloidal) soils are primarily due to surface tension forces and associated negative pore-water pressures. In their analysis, Haines and Fisher considered the case of "ideal" soils composed of uniform sized spherical particles arranged in close-packed and open-packed arrays. For these simple systems they were able to calculate the cohesive stresses as a function of soil water content and particle size.

Some very interesting observations from the point of view of scaling can be made from their results. The first is that for any given soil water content and particle packing arrangement, the pore-water pressure deficiency p is proportional to the factor σ/r, where σ is surface tension and r is the particle radius. This is exactly what would have been predicted from the Miller and Miller (1956) scaling theory. The second observation is that the cohesive stress τ is also proportional to σ/r, which indicates that τ scales in the same manner as p. A corollary to these two observations is that the dimensionless ratio τ/p depends only on water content for similar soils of the Haines-Fisher type.

For a long time the results of Haines and Fisher received only limited attention in the soil physics community. In the 1960s their theory was tested in a series of experiments using glass bead systems (Vomocil & Waldron, 1962). Results were rather disappointing in that measured values of tensile strength were considerably lower than those predicted by the Haines-Fisher theory. This was attributed primarily to the possibility that the soil water matric potential, which was used as an estimator of the pore-water pressure, contained a significant "adsorption" component in addition to the capillary pressure component. In such a situation the estimated pore-water pressure would be too negative, with the result that the calculated soil tensile strength would be too high.

The problem was reexamined 20 yr later by Snyder and Miller (1985). They suggested that part of the discrepancy between measured and estimated values of tensile strength in the glass bead experiments of Vomocil and Waldron (1962) was due to the fact that the Haines-Fisher theory did not take into account the concentration of tensile stresses around flaws or "cracks" in the particle packing arrangement. Experimental difficulties associated with slow water equilibration in systems with unimodal pore-size distributions (i.e., glass bead systems) were also proposed as a factor.

In their analysis, Snyder and Miller (1985) generalized the Haines and Fisher theory to include granular systems of any geometrical configuration which retained the property of being similar according to the rules defined by Miller and Miller (1956). Their results preserved the same scaling property as those obtained by Haines and Fisher, namely that the ratio τ/p is only a function of the degree of pore saturation for granular similar media. The theory was observed to agree very well with measured tensile strength values of granular soils (sands and silts) with low colloid content. When used as a first approximation for moist aggregated soils with relatively high clay content, the theory provided a way of normalizing the experimental data in a way that greatly reduced the variability between different soils.

These results for soil tensile strength, and the fact that the underlying principles are fundamentally related to the Miller and Miller scaling theory, raises the intriguing question as to whether scaling theory can be extended to soils subjected to more complex stress systems than simple tension. The question is of relevance to important soil processes such as compaction, reaction to tillage implements, and mechanical impedance to penetration. If shown to be relevant even as a first approximation for a group of soils, such a theory would allow characterization of both hydraulic and mechanical strength properties from similar sets of measurements.

The data are currently not available to evaluate such a possibility completely. In view of this, the present chapter will be limited to describing the physical situation in an unsaturated granular soil under an applied mechanical load as clearly and accurately as possible. The underlying mathematical relationships will be formulated and subjected to dimensional analysis, and resulting implications for theoretical and experimental aspects of soil mechanical behavior will be discussed. Supporting experimental data will be cited whenever available for a given aspect of the theory.

I hope this chapter will illustrate the need and many opportunities for further research on a subject that has received only limited attention in the field of soil physics.

DESCRIPTION OF THE SYSTEM

The soil system to be considered is essentially the same as described by Miller and Miller (1956), where pore liquid behavior is governed entirely by gravity and surface tension effects. The pore liquid phase is allowed to be either in equilibrium or in a state of "creeping flow" where inertial forces are negligible. In addition, a static load is applied at the boundary of the soil system, as shown in Fig. 7–1. The particle system (solid phase) is considered to be in static equilibrium.

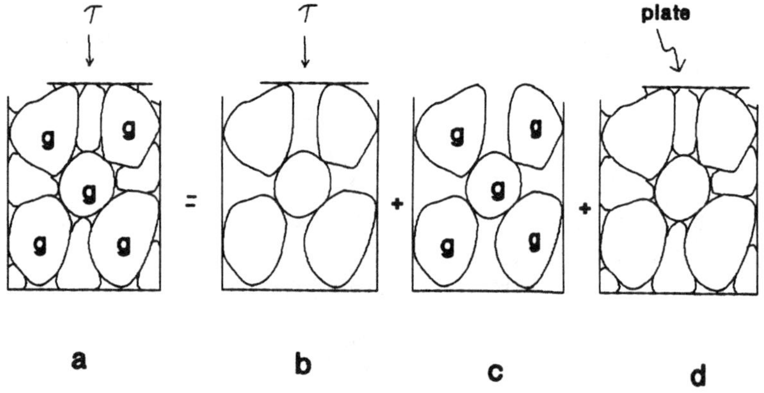

Fig. 7–1. Representation of the state of stress (*a*) in a granular soil as the linear superposition of, (*b*) an applied stress system, (*c*) a gravitational force system, and (*d*) the liquid phase stress system due to the presence of a gas-liquid interface.

It is assumed that the only forces acting on soil particles are those caused by gravity, boundary forces, surface tension, and stresses in the pore phases. Forces associated with electrostatic (double layer) interactions and chemical bonding at interparticle points of contact are assumed negligible. Slippage at the interparticle contact points is assumed to be governed by Coulombic friction interactions. The interparticle contact points are treated as point contacts. These assumptions have been found to be reasonable approximations for recently disturbed granular soils such as sands and silts (Skempton, 1960; Lamb & Whitman, 1969; Wu, 1976). Individual soil particles are considered as essentially rigid bodies, with high enough bulk and shear modulii that "small displacement" theory of elasticity can be applied to any given particle. The assumption of small displacement elasticity is common in rock mechanics (Hiramatsu & Oka, 1966); thus, it should apply reasonably well to individual granules in mineral soils.

Following the principle of effective stress first postulated by Terzaghi (Skempton, 1960), it will be assumed the forces that actually cause "noticeable" deformation of a granular soil particle framework are those at interparticle contact points within the medium. Stresses within the particles themselves are assumed to have only a minor effect on soil deformation, because the particles rarely break and only deform negligibly and reversibly under applied loads. "Noticeable" soil deformation will happen when the intergranular force system is such that interparticle separation (tensile failure), slippage (shear) or rolling (also a form of shear failure) can occur. This type of deformation is essentially irreversible (i.e., plastic); once the particles have been forced to move relative to each other it is very unlikely they will return to their original positions upon unloading.

Due to the essentially undeformable nature of the individual soil particles, soil deformation under applied loads is often very small prior to the point of irreversible failure. This is especially true in the case of granular soils. The present chapter will be limited to soils that are within the "small strain" range. This restriction is necessary for the superposition of forces, which will be described later, to be valid. It need not be considered too confining a restriction, however, because in principle it still allows analysis of soils that are stressed to the limit just prior to plastic deformation. This "limit stress" is precisely what is pursued in fields such as critical state soil mechanics (Hettiaratchi & O'Callaghan, 1984).

The intergranular force system is usually treated in soil mechanics as a stress field. This approach, in essence, requires that the (discrete) interparticle forces acting over any given "representative" cross-sectional area of the soil medium be added and then divided by the area to obtain "stresses" that are assumed to be continuously differentiable spatial functions of the material. The concept of an intergranular stress field is mathematically convenient because it allows access to the powerful methods of continuum mechanics. Its nature as a smoothed approximation of a discrete force system, however, can result in loss of detail that may be important in explaining differences in soil behavior. Because of this, the approach used here will be to consider the intergranular force system, rather than an intergranular stress field, as the fundamental soil variable. We keep in mind, however, that, because of size scale differences between external loading systems and a soil particle, the quantities that can most conveniently be measured are stresses—hence the need to ultimately be able to relate between (microscopic) forces and (macroscopic) stresses.

When a load is applied at the boundary of a soil system, the mode in which it is transmitted from particle to particle is essentially a probabilistic process. This is because it depends on the largely random orientation and position of soil particles relative to each other. One can only speak of the probability of occurrence of a given interparticle force at a given spatial position within a granular medium. This has been recognized by a number of authors (Smoltczyk, 1967; Sergeev, 1969; Harr, 1977) who have used probability theory to explain the distribution of stresses in soils under applied loads. The fundamental assumption behind these approaches is that interparticle forces in soils are transmitted from particle to particle according to a "random walk" Markov chain process. Although this assumption seems rather simplistic, it has been used to calculate stress distributions that are in good agreement with those obtained by more elaborate continuum methods (Harr, 1977). Furthermore, the approach is fundamentally based on the microscopic structure of a granular medium. It can, thus, be used to obtain much information on the scaling properties of intergranular force transmission in soils. Scaling properties derived from the Markov assumption will have important implications in this chapter.

SUPERPOSITION ASSUMPTION

A simultaneous analysis of all the interacting forces in an externally loaded unsaturated soil, while physically the most general approach, would be very complex. If we restrict our analysis to the "small strain" condition discussed previously, however, the problem can be greatly simplified by using a superposition procedure. This involves dividing a problem into simpler, independent problems that can be solved, then adding or "superimposing" the solutions to obtain the final answer. Superposition techniques are commonly used for solving complex stress-distribution problems in theory of elasticity (Timoshenko & Goodier, 1970).

In our situation we assume that interparticle forces in an unsaturated granular soil, with particles in static equilibrium, can be represented as the linear superposition of interparticle forces in the three independent equilibrium systems shown in Fig. 7–1. The three systems are: (Fig. 7–1b) the particle system under the action of applied (boundary) forces only; (Fig. 7–1c) the particle system under the action of gravity only; and (Fig. 7–1d) the particle framework under the action of the liquid phase. Each of these components will be analyzed in terms of the scaling properties of its effects on interparticle contact forces. Results will then be superimposed and inspected as a whole.

To give an idea of the reason for the particular choice of component problems, we appeal to common observations on the behavior of granular systems. Consider an unsaturated granular soil in an open container on a table top. A loaded plate is placed on the soil surface. We wait for the soil particle framework to reach equilibrium under this load. The system in equilibrium is shown in Fig. 7–1a.

All external boundary forces, except those absolutely necessary to maintain the liquid phase in an unaltered state and to sustain the system against gravity, are now removed. In our example of the soil on the table, this corresponds to removing the applied load from the plate without removing anything else, including the plate. Leaving the (weightless) plate in place is necessary for the geometry of the liquid phase at the soil-plate interface to remain unaltered. Leaving the container walls and table in place is required to prevent gravity from accelerating the system in space. Boundary forces remain acting on the system, but they are only those necessary to prevent the liquid phase from moving the boundaries and those associated with sustaining the soil against gravity.

For reasons discussed previously, removal of the load from the plate will result in negligible deformation of the soil particle system; i.e., it will remain in static equilibrium in essentially the same configuration it had before the load was removed. Because the solid phase geometry is unchanged, neither is the state of the liquid phase since, in granular soils, it will depend only on the geometry of the (unaltered) particle framework. Thus, the unloaded soil-liquid-gas system is still in equilibrium in the same geometrical configuration it had before unloading. The implication is that a boundary force system acting on a soil in equilibrium has been removed such that the remaining system is still in equilibrium, which indicates the removed force system together with the interparticle contact forces that it caused can be treated as an independent equilibrium system as represented in Fig. 7–1b.

We next allow our soil to completely dry, so that forces due to the liquid phase (including the intergranular and boundary forces induced by it) are removed from the system. Common experience has shown only small deformation upon drying occurs in many granular soils, so the dry particle system can be considered to be in equilibrium in the same geometrical configuration it had in the soil before drying. Again, we have removed a force system from a soil in equilibrium to obtain another system that is still in equilibrium. The removed force system (liquid phase effects) can thus be considered an independent force system as illustrated in Fig. 7–1d. Note that gravitational effects on the liquid phase need to be included in this force system.

Our final equilibrium system is the particle framework within its confining boundaries, under the action of gravity only. This is represented in Fig. 7–1c.

To keep our discussion as simple and clear as possible, an important consideration has been omitted until now. This is the effect of the gaseous phase in the soil pores. It is tacitly implied in Fig. 7–1 that the pore gas pressure is in equilibrium with the surrounding atmospheric pressure, which allows taking the "gauge" pore-gas pressure as zero. This need not always be the case, however, as if for example the soil were isolated from the atmosphere by an impermeable membrane. How is the system affected if we impose a pressure difference (which we simply call the gas pressure) between the atmospheric and pore gas phases?

The answer is relatively straightforward. We first note that any change in pore gas pressure will result in an *equal* pressure change of the liquid phase, without in any way affecting its geometry provided the liquid is incompressible. Because of this uniform change in pressure of all pore phases, the pressure everywhere on the surface of a soil particle will change by the same amount independently of which phase (gas or liquid) is touching the particle at a given point. The pressure change will tend to compress or expand the particle (a negligible amount), but because it is a spatially uniform pressure change the particle as a whole will not be "pushed" or rotated in any direction. Thus, *the net effect of a gas pressure change on the interparticle contact forces (forces with which particles "push" each other) is zero*. This property of a uniform fluid pressure change in a granular medium underlies Terzaghi's principle of effective stress (Skempton, 1960).

The overall pore-fluid pressure change associated with a change in gas pressure will also cause a uniform pressure change on the inside of the boundaries of the soil medium. Here it is assumed that the points of contact of the soil particles with the boundary are small enough that for practical purposes the entire surface area of the boundary is in contact with pore fluids (Skempton, 1960). The fluid pressure change will act against the normal components of the external loading stresses acting on the boundary. Thus, the "effective" *normal* components of the boundary stresses (the normal stress components that the particle matrix actually "feels" as a result of the gas pressure change) will be the difference between the actual normal stress components and the pore gas pressure. The *shear stress* components of the applied stress, on the other hand, will be unaffected by the pressure change because the latter is incapable of withstanding shear. Taking these factors into consideration, the "effective" boundary stress can be written in tensorial form as

$$\boldsymbol{\tau}' = \boldsymbol{\tau} - \mathbf{u} \cdot \mathbf{1} \qquad [1]$$

where $\boldsymbol{\tau}'$ is the effective boundary stress tensor acting at a point, $\boldsymbol{\tau}$ is the total (actual) boundary stress tensor, u is the pore gas pressure, and $\mathbf{1}$ is the unit tensor. The effective applied stress $\boldsymbol{\tau}'$ of Eq. [1] is the correct stress to use as the "applied stress component" (Fig. 7–1b) in our superposition problem. The liquid phase and gravitational effects (Fig. 7–1d and 7–1c, respectively) remain the same as before, with the understanding that the liquid pressure to be considered is its *pressure deficiency* p relative to the gas phase.

In the case of soils with only one pore fluid (such as saturated soils), the multiphase component Fig. 7–1d is nonexistent. The problem in Fig.

7–1 is then reduced to the force components of Fig. 7–1b and Fig. 7–1c, with u of Eq. [1] being the pressure of the pore fluid in question. Equation [1] in this case simply represents Terzaghi's principle of effective stress for single-fluid soils, which states that the mechanical behavior of a single-fluid soil is a function of $\boldsymbol{\tau}'$ (Skempton, 1960).

Having stated the nature of our problem, we now proceed to analyzing each of the force components in Fig. 7–1 in turn.

FORCE SYSTEM 1: EQUILIBRIUM UNDER EFFECTIVE APPLIED STRESSES

The conditions for static equilibrium of a particle framework (Fig. 7–2) are that the sums of forces and first moments (torques) acting on the system must be zero (Meriam, 1975). These conditions must apply simultaneously to the total particle framework (it cannot accelerate either linearly or rotationally) and to each individual particle within the framework. For a given particle within the system, this implies that

$$\sum_{k=1}^{K} \boldsymbol{f}_k = 0 \qquad [2]$$

and

$$\sum_{k=1}^{K} (\boldsymbol{f}_k \times \boldsymbol{r}_{km}) = 0 \qquad [3]$$

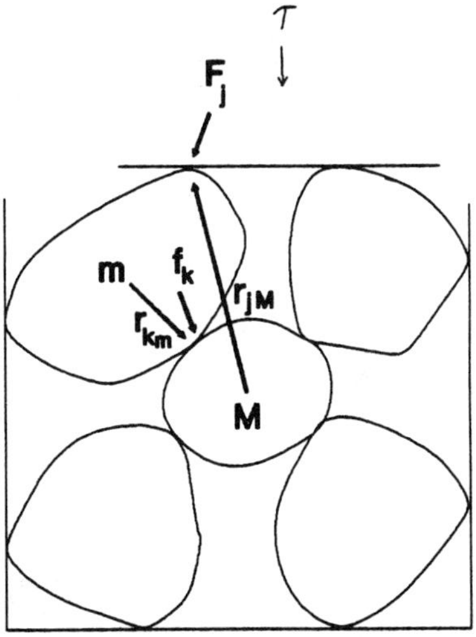

Fig. 7–2. Boundary and interparticle forces acting on an externally loaded particle framework in the absence of gravity and a gas-liquid interface.

where $\boldsymbol{f}_k$ is the interparticle force at the kth contact point of the particle, and the vector $\boldsymbol{r}_{km}$ indicates the position of the contact point k relative to the center of mass m of the particle. The upper case symbol K in Eq. [2] and [3] represents the coordination number of the particle, or the number of contact points associated with it.

Equations similar to Eq. [2] and [3] govern the relationships between the boundary forces at equilibrium

$$\sum_{j=1}^{J} \boldsymbol{F}_j = 0 \qquad [4]$$

and

$$\sum_{j=1}^{J} (\boldsymbol{F}_j \times \boldsymbol{r}_{jM}) = 0 \qquad [5]$$

The variables $\boldsymbol{F}_j$ indicate the boundary forces at points $j = 1, 2, 3 \ldots J$, respectively, with the symbol $\boldsymbol{r}_{jM}$ representing the corresponding position vector relative to the center of mass M of the system.

It should be noted that the equilibrium conditions given in Eq. [2] and [3] for individual particles are not independent of the boundary conditions in Eq. [4] and [5] because points on the boundary are also points on individual particles.

An important property of the system of Eq. [2] to [5] is that all forces can be divided by a constant scale factor a_f and all position vectors by a constant "microscopic" length scale factor a_l to obtain the dimensionless equations

$$\sum_{k=1}^{K} \boldsymbol{f}_k^* = 0 \qquad [6]$$

$$\sum_{k=1}^{K} (\boldsymbol{f}_k^* \times \boldsymbol{r}_{km}^*) = 0 \qquad [7]$$

$$\sum_{j=1}^{J} \boldsymbol{F}_j^* = 0 \qquad [8]$$

and

$$\sum_{j=1}^{J} (\boldsymbol{F}_j^* \times \boldsymbol{r}_{jM}^*) = 0 \qquad [9]$$

where the new dimensionless variables are defined as

$$\boldsymbol{f}_k^* = \boldsymbol{f}_k/\alpha_f \tag{10}$$

$$\boldsymbol{F}_j^* = \boldsymbol{F}_j/\alpha_f \tag{11}$$

$$\boldsymbol{r}_{km}^* = \boldsymbol{r}_{km}/\alpha_l \tag{12}$$

$$\boldsymbol{r}_{jM}^* = \boldsymbol{r}_{jM}/\alpha_l \tag{13}$$

The units of α_f and α_l are force and length, respectively. The scaling properties of the force balance equations Eq. [2] to [13] imply that, as long as geometrical similitude is maintained with respect to the particle matrix and the position and direction of applied boundary forces, the reduced forces at any given reduced position are always the same regardless of the scale factors α_f and α_l. A requirement is that the elastic properties of all particles must be the same, particularly in the case of statically indeterminate particle frameworks as will be discussed later. Similar media loaded in such a way that the force scaling criteria Eq. [10] to [13] are always maintained will be termed *similarly loaded* similar media.

The interparticle contact force $\boldsymbol{f}_k$ at any given point within a loaded granular medium will be determined by the sum of the effects of the discrete boundary forces $\boldsymbol{F}_j$ acting on the medium. The contribution of each of the boundary forces to $\boldsymbol{f}_k$ can be written

$$\boldsymbol{f}_{kj} = \boldsymbol{A}_{kj} \cdot \boldsymbol{F}_j \tag{14}$$

where $\boldsymbol{f}_{kj}$ is the component of $\boldsymbol{f}_k$ contributed by $\boldsymbol{F}_j$, and $\boldsymbol{A}_{kj}$ is a dimensionless operator that defines the relationship between the two. In computational form, $\boldsymbol{A}_{kj}$ can be considered as a matrix which, when multiplied by the vector (column matrix) $\boldsymbol{F}_j$, produces another vector $\boldsymbol{f}_{kj}$. Because $\boldsymbol{f}_k$ is the sum of the effects of all the boundary forces $\boldsymbol{F}_j$ we can write it, with the help of Eq. [14], as

$$\boldsymbol{f}_k = \sum_{j=1}^{J} \boldsymbol{f}_{kj} = \sum_{j=1}^{J} (\boldsymbol{A}_{kj} \cdot \boldsymbol{F}_j) \tag{15}$$

The operator $\boldsymbol{A}_{kj}$ has some important properties. We first note that, because of the scaling relations Eq. [6] to [13], the relation between $\boldsymbol{f}_{kj}$ and $\boldsymbol{F}_j$ in Eq. [14] must be independent of changes in the length scale α_l of granular similar media. This implies that $\boldsymbol{A}_{kj}$ is also independent of length scale. It is only a function of the reduced position vectors $\boldsymbol{r}_{km}^*$ and $\boldsymbol{r}_{jM}^*$ of $\boldsymbol{f}_{kj}$ and $\boldsymbol{F}_j$, respectively. We also note from the similarity relations Eq. [6] to [13] that a proportional change α_f in all the boundary forces $\boldsymbol{F}_j$ in similarly loaded similar media will cause all the interparticle forces $\boldsymbol{f}_k$ (or $\boldsymbol{f}_{kj}$) to change in the same proportion. Applying this condition to Eq. [14] or [15] shows that $\boldsymbol{A}_{kj}$ has the linear property

$$\boldsymbol{A}_{kj} \cdot (\alpha_f \boldsymbol{F}_j) = \alpha_f (\boldsymbol{A}_{kj} \cdot \boldsymbol{F}_j) \qquad [16]$$

Equations [15] and [16] will be of considerable use later in this chapter.

Questions have been raised on the possibility of an analysis of granular systems such as that embodied in Eq. [2] to [16], on the grounds that interparticle friction forces render the systems indeterminate (Brown & Richards, 1970). It is argued that, because frictional forces can assume any value ranging between zero and a limiting value just prior to that causing slippage failure, the systems must be indeterminate. In analyzing this argument, however, it is important to note that for a granular system in equilibrium, friction forces at intergranular contact points are simply *reactions* of the material against the tangential (slippage) components of the intergranular contact forces induced by applied loads. Because the system is in static equilibrium, friction forces must be opposite in sign and equal in magnitude to the slippage forces, and as such are completely and uniquely determined by the direction and magnitude of the interparticle contact forces relative to the particle surfaces. Indeterminacy due to friction is thus not a problem.

A problem of static indeterminacy, however, can arise in situations where the number of supports or constraints (in this case interparticle and boundary contact points) are more than those absolutely necessary to ensure static equilibrium. The problem can be simply visualized by comparing the equilibrium state of a three-legged stand on a floor with that of a five-legged stand. Because three legs are the minimum required to ensure that the stand will not tip over, one can with relative ease calculate the load on each leg using only simple laws of statics similar to those embodied in Eq. [3] to [14] (Meriam, 1975). The three-legged stand is an example of a statically determinate structure. On the other hand, the five-legged stand represents a statically indeterminate structure because—provided the legs are of the same length—there are more than enough legs to keep the stand from tipping. In mathematical terms, we now have more unknown variables (stand legs) than equations (simple laws of statics). To find the load on each leg we would have to combine the principles of statics with more complex methods, such as theory of elasticity. A similar problem exists with calculating intergranular contact forces in statically indeterminate particle systems using only the laws of statics embodied in Eq. [2] to [16]. If we assume the elastic properties of all soil particles are the same, however, the *scaling* properties of Eq. [2] to [16] are still preserved. In other words, the invariance of reduced forces at reduced positions is still maintained upon changes in force or length scale. This principle is important in scale model tests of statically indeterminate structures in engineering practice (David & Nolle, 1982). Because we are only interested in the scaling properties of granular systems under loads, we consider our problem resolved. This is provided, of course, that the assumption of uniform elastic properties of soil particles is a reasonable one.

Another difficulty with attempting to apply scaling theory to granular media is the fact that particles in "similar media" will rarely, if ever, be arranged in exactly the same manner. As mentioned previously, one can only speak of the probability of occurrence of different possible particle arrangements and resulting modes of force transmission through the particle framework. Harr (1977) has suggested that, because the possible number of particle arrangements is so large, the probabilities of different modes of force transmission between two points in a granular medium will tend toward a normal distribution. If we assume that the probability distribution of different possible particle arrangements and, hence, modes of force transmission in statistical ensembles of n particles is independent of particle size (α_l), then it follows from previous discussion that the probability distribution of scaled forces at given scaled positions should be invariant for similarly loaded similar media. We thus assume that, "on the average," our scaling properties will hold.

The analysis until now has been limited to the effects of *point forces* ($\boldsymbol{F}_j$) at the surface of a granular medium on the interparticle forces in its interior. Surface forces, however, are usually applied as distributed loads, which can only be measured as stresses. Thus, it is important to be able to relate surface stresses to interparticle forces within the medium. Following our concern with scale effects, we ask ourselves how the relationships are affected by changes in the scale of the applied stress system and of the particle framework.

An additional consideration arises at this point in that now *two* length scales become important. One is the microscopic characteristic length α_l with which we have been dealing, and the other is a macroscopic characteristic length L associated with the dimensions of the applied stress system. For example, consider the simple case of a tractor tire pressing against the soil. One can vary the tire size (L), the soil particle size (α_l) or both independently. It is important to know how interparticle forces are affected by different combinations of such scale changes.

Three simple cases, from which all combinations of changes in L and α_l can be constructed, are illustrated in Fig. 7–3. The first case (Fig. 7–3a,b) compares two loaded soil systems that are exact geometrical analogues of each other. Such a situation requires that all changes in α_l and L must be made subject to the constraint that α_l/L = constant. Because of this constraint only one of the two length scales is independent; we choose the microscopic α_l as the independent scale. The second case (Fig. 7–3b,c) involves a change in α_l whereas L remains constant. The third case (Fig. 7–3a,c) compares changes in L at constant α_l. The geometrical shape of the loaded surface is similar for all soils, and a condition of "similar loading" (to be defined later) is imposed.

We first consider Case 1 where both L and α_l are required to vary in the same proportion, as illustrated in Fig. (7–3a,b). To analyze the problem, a functional relationship is needed between the macroscopically continuous applied stress $\boldsymbol{\tau}'$ and the resulting microscopic (discrete) forces $\boldsymbol{F}_j$ induced at the $j = 1, 2 \ldots J$ contact points between surface particles and the loaded boundary. We achieve this by distributing a fraction of the total surface load to each of the J surface particles with the function

$$\boldsymbol{F}_j = \boldsymbol{\beta}_j^* \cdot \int_A \boldsymbol{\tau}' \cdot \boldsymbol{n}\ dA \qquad [17]$$

where $\boldsymbol{F}_j$ is the force on the jth boundary particle, $\boldsymbol{n}$ is the unit vector normal to any infinitesimal element area dA on the loaded surface, and $\boldsymbol{\beta}_j^*$ is a dimensionless vector operator that will be defined shortly. The product of the stress tensor $\boldsymbol{\tau}'$ and the vector $\boldsymbol{n}\ dA$ yields the force acting on the element dA (Symon, 1971). The integral of all such forces over the entire loaded surface area A in Eq. [17] then yields the total force vector acting on the surface. Each particle on the surface will receive a fraction (varying between zero and one) of this total load. The fraction of the load transmitted to the jth particle, as well its direction, is indicated by the operator $\boldsymbol{\beta}_j^*$ in Eq. [17]. It is assumed that $\boldsymbol{\beta}_j^*$ is independent of microscopic scale α_l for systems such as in Fig. 7–3a,b, which are exact scale models of each other. In all such systems, the number of particles and their geometrical arrangement is identical independently of scale. Thus, the probability that a given fraction of the load will be distributed to a given particle

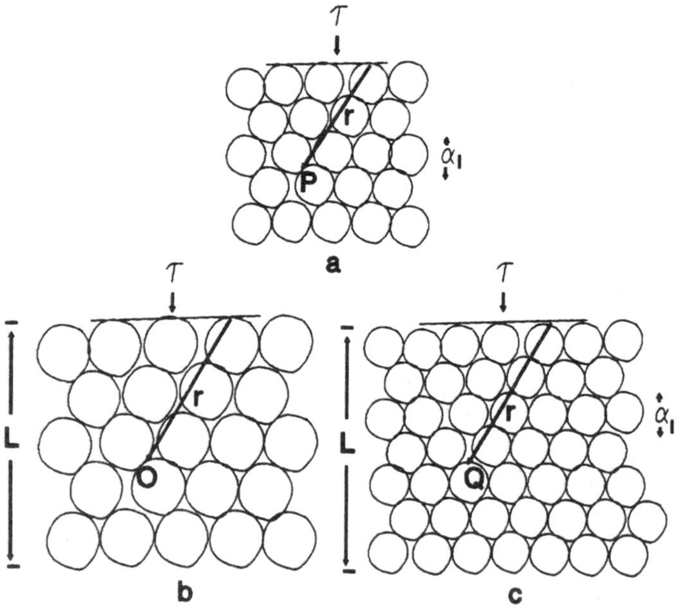

Fig. 7–3. Distributed surface loads (stresses) on granular systems under different combinations of microscopic and macroscopic changes in scale.

should be constant. This of course assumes a condition of "similar loading" such that the direction and *relative* magnitude of the applied stress at any given reduced position $\boldsymbol{r}/\alpha_l$, on the soil surface is constant. This condition can be written in terms of the gradient of the applied stress over the soil surface as

$$\overline{\boldsymbol{V}}_m^* \boldsymbol{\tau}'^* = (\alpha_l \overline{\boldsymbol{V}})(\boldsymbol{\tau}'/\alpha_\tau) = f(\boldsymbol{r}/\alpha_l) \qquad [18]$$

where the symbol $\overline{V}$ represents the gradient of the stress $\boldsymbol{\tau}'$ over the loaded soil surface, $\boldsymbol{V}_m^*$ is the microscopically reduced gradient $\alpha_l \overline{\boldsymbol{V}}$, α_l has its usual meaning, α_τ is a scalar factor with units of stress, and $f(\boldsymbol{r}/\alpha_l)$ denotes a unique function f of the reduced position $\boldsymbol{r}/\alpha_l$. The reduced stress $\boldsymbol{\tau}'^*$ $(= \boldsymbol{\tau}'/\alpha_l)$can be considered as a "relative stress tensor," which indicates the direction and relative magnitude of the applied stress at any reduced position on the soil surface. The stress scale factor α_τ has a meaning analogous to the force scale factor α_f of Eq. [2] to [9]. It represents the scale or "mean magnitude" of the applied stress system. When multiplied by the relative stress $\boldsymbol{\tau}'^*$ at any point it yields the actual stress $\boldsymbol{\tau}'$.

With use of the substitutions $A^* = A/\alpha_l^2$ and $\boldsymbol{\tau}'^* = \boldsymbol{\tau}'/\alpha_\tau$, Eq. [17] can be written

$$\boldsymbol{F}_j = \alpha_\tau \alpha_l^2 (\boldsymbol{\beta}_j^* \cdot \int_{A^*}^{\tau *} \cdot \boldsymbol{n}\, dA^*) \qquad [19]$$

This equation groups the surface contact forces into a dimensionless geometrical component (the factor in parentheses) and the scale component $\alpha_\tau \alpha_l^2$ which accounts for changes in particle size and stress magnitude.

We are now ready to determine the effects of these boundary conditions on the interparticle forces $\boldsymbol{f}_k$ within the granular medium. As mentioned previously this is simply a matter of adding the effects of all the discrete boundary forces $\boldsymbol{F}_j$. Substituting Eq. [19] into Eq. [15] accordingly, and making use of the identity Eq. [16] we obtain

$$f_k = \alpha_\tau \alpha_l^2 \sum_{j=1}^{J} \left\{ \boldsymbol{A}_{kj} \cdot [\boldsymbol{\beta}_j^* \cdot \int_{A^*} (\boldsymbol{\tau}'^* \cdot \boldsymbol{n} \;\; dA^*)] \right\} \qquad [20]$$

This equation is simpler that it seems. Note that the operations inside the square brackets of Eq. [20] yield a dimensionless vector, because they represent the dimensionless vector operator $\boldsymbol{A}_{kj}$ operating on the dimensionless vector

$$\boldsymbol{\beta}_j^* \cdot \int_{A^*} \boldsymbol{\tau}'^* \cdot \boldsymbol{n}\, dA^*$$

defined in Eq. [19]. The summation of these vectors as indicated in Eq.

[20] is also a dimensionless vector, which we denominate $f^*_{k\tau}$. Substituting this into Eq. [20] yields

$$f_k = (\alpha_\tau \, \alpha_l^2) f^*_{k\tau} \qquad [21]$$

This simple result states that, as long as strict geometrical similitude of the soil and loading system is maintained, the intergranular force at any given reduced position r/α_l within the medium is directly proportional to the scalar coefficient $(\alpha_\tau \alpha_l^2)$. The direction of f_k and its relative magnitude with respect to the interparticle forces in other parts of the soil medium, are indicated by the dimensionless vector $f^*_{k\tau}$.

We now turn our attention to the second scaling case of Fig. (7–3a, c) where the macroscopic scale L is allowed to vary but the microscopic scale α_l is held constant. This situation is more difficult to analyze than the previous case because the condition of strict similarity (which requires a constant number of soil particles) is not met. The problem will be addressed by using the probabilistic approaches discussed previously.

The rationale for using probability theory to describe interparticle force transmission in soils is illustrated in Fig. 7–4 for a two-dimensional stack of particles. A point force is applied to a particle on the system boundary. This force is transmitted to the two particles directly below it, such that each of the latter will now (most probably) bear half the load of the original particle. We note that the process occurs in discrete "jumps" over finite distances Δz and Δx, and that the load is distributed laterally as well as downward during the process. We also note that the values of Δz and Δx *depend on the structure of the granular medium*. The two par-

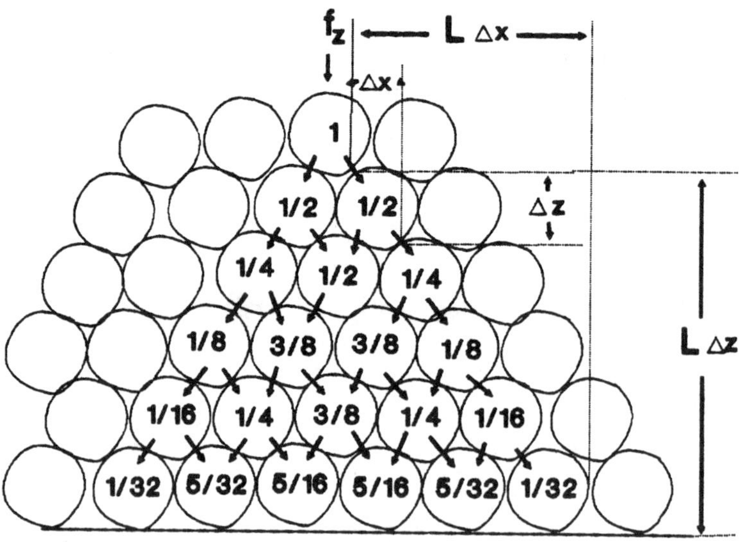

Fig. 7–4. Visualization of intergranular force transmission in soils as a random walk process.

ticles that received a fraction the original force now in turn transmit their fractions to neighboring particles and the sequence continues in a chain reaction. The fraction of the surface load that will be transmitted to each particle by the end of the (instantaneous) process is illustrated in Fig. 7–4 with the number inside the particle.

Now suppose that instead of a force on the top particle in Fig. 7–4, we have the famous "drunk man" of probability theory who decides to take a random walk from particle to particle. In each step from one particle to the next, the man is allowed to go either to his right or to his left, provided he always goes a step down for every lateral step. Probability theory can be used to estimate the probability that the man will visit any given particle during his walk. It turns out (Harr, 1977) that the probabilities are the same as the force fractions we calculated in our simple force transmission problem. This is important, for it means that we have at our disposal some very powerful mathematical methods of probability theory with which to simulate the interparticle force transmission process.

The mathematical description of the random walk process we have just described begins with the simple recurrence relation (Harr, 1977)

$$f_z(x, z + \Delta z) = 1/2[f_z(x - \Delta x, z) + f_z(x + \Delta x, z)] \qquad [22]$$

This equation states that the probability $f_z(x, z + \Delta z)$ of transmission of the z-component of a surface force to a position $(x, z + \Delta z)$ in a granular medium is equal to the average of the respective probabilities $f_z(x - \Delta x, z)$ and $f_z(x + \Delta x, z)$ that the force will be felt at the immediately antecedent neighboring positions $(x - \Delta x, z)$ and $(x + \Delta x, z)$. The probability $f_z(x, z + \Delta z)$ can also be interpreted as the most probable fraction of the load on a surface particle that will be transmitted to another particle at position $(x, z + \Delta z)$ relative to it. The latter interpretation is perhaps the most useful within the context of force transmission in soils.

Subtracting the probability $f_z(x, z)$ from both sides of Eq. [22], dividing by Δz and performing some algebraic manipulation yields (Harr, 1977, p. 126).

$$\Delta f_z/\Delta z = (-1/2\, \Delta x^2/\Delta z)[\Delta(\Delta f_z/\Delta x)/\Delta x] \qquad [23]$$

This equation describes the two-dimensional "diffusion" of z-component f_z of a force as it "jumps" in discrete steps between interparticle contact points at average distances Δx and Δz away from each other as we illustrated in Fig. 7–4. The *force diffusion coefficient* $-1/2\, \Delta x^2/\Delta z$ is related to the average lateral distance Δx over which the force is transmitted every time it is transmitted a vertical distance Δz. If we view the system from far enough away that the microscopic "force jump" distances Δx and Δz tend to zero, our difference equation can be written as a linear partial differential equation

$$\delta f/\delta z = D\, \delta^2 f/\delta x^2 \qquad [24]$$

where

$$D = -1/2\, \Delta x^2 / \Delta z \tag{25}$$

Notice that D is essentially a *geometrical* property of the medium, which indicates how the interparticle contact forces are to be distributed in mutually perpendicular directions; it is not part of the force transmission process itself. To emphasize this property of D and the fact that it depends very markedly on microscopic structural geometry, we leave D expressed in difference rather than differential form.

Equations [24] and [25] are written in two dimensions. However, following Harr (1977), the two-dimensional solution $f_z(x, z)$ can be extended to three dimensions by remembering its analogy to the random-walk *probability* that f_z would be found at any given location in the x–z plane. If soil properties (i.e., D) are symmetrical around the z axis, the same probability distribution would be found for a two-dimensional problem in the y–z plane. The three-dimensional probability can then be treated as the joint probability (i.e., the product) of the mutually independent two-dimensional probabilities. Thus, we have

$$f_z(x, y, z) = f_z(x, z) f_z(y, z) \tag{26}$$

Because of this property, we need only understand the behavior of two-dimensional systems to predict the behavior of three-dimensional situations.

Note the direction z of Eq. [24] and [25] represents only one of the three mutually orthogonal component directions of a surface force vector. Hence, in actuality we would have *three* differential equations similar to Eq. [24] and [25]. Still another set of equations would be needed to describe the diffusion of moments. Because the equations are similar in form to Eq. [24] and [25], however, we only consider the scaling properties of the latter.

We can evaluate the influence of microscopic scale α_l and force scale α_f on the system described in Eq. [24] and [25] by introducing the dimensionless variables

$$x_m^* = x/\alpha_l,\ z_m^* = z/\alpha_l,\ f_z^* = f_z/\alpha_f$$

where the subscript m indicates *microscopic* scaling, and substituting them into Eq. [24] and [25] to obtain

$$\delta f_z^* / \delta z_m^* = D^* \delta^2 f_z^* / \delta x_m^{*2} \tag{27}$$

where

$$D^* = -1/2\, \Delta x_m^{*2} / \Delta z_m^* \tag{28}$$

This implies that the reduced force function $f_z^*(x_m^*, z_m^*)$ is invariant under changes in microscopic scale (with macroscopic scale L changing proportionally) and conditions of similar loading. This, of course, is in agreement—as it should be—with the conclusions we drew from Eq. [2] to [13] for systems that are exact geometrical analogues of each other.

With certain assumptions, Eq. [24] and [25] also allow evaluating the effects of changes in macroscopic characteristic length L at constant α_l. Let us suppose that as indicated in Fig. 7–3a and 7–3c, the spatial dimension (macroscopic length L) of the applied stress system is changed while the microscopic scale α_l remains constant. The system is again "similarly loaded," but because this time we are interested in changes in *macroscopic* scale, the gradient and position of the applied surface stresses are required to scale in terms of L as

$$\overline{\boldsymbol{V}}_M^* \boldsymbol{\tau}'^* = (L\overline{\boldsymbol{V}})(\boldsymbol{\tau}'/\alpha_\tau) = f(\boldsymbol{r}/L) \qquad [29]$$

where $\boldsymbol{V}_M^*$ is the macroscopically reduced gradient $L\overline{\boldsymbol{V}}$ and $f(\mathbf{r}/L)$ denotes a function f of the macroscopically reduced position $\boldsymbol{r}/L$. In addition, we assume the soil is spatially isotropic, so that a given stress placed anywhere on the soil surface will always yield the same distribution of discrete surface forces $\boldsymbol{F}_j$ per unit area. Under this condition and Eq. [29], the surface force $\boldsymbol{F}_j$ at any given macroscopically scaled position $\boldsymbol{r}/L$ on the surface can be assumed to be constant for macroscopically similar systems. We note that this invariance of $\boldsymbol{F}_j$ is only with respect to changes in the macroscopic scale L. If we varied the microscopic scale α_l and the stress scale factor α_τ in addition to L, Eq. [19] combined with Eq. [29] tells us that $\boldsymbol{F}_j$ at any reduced position $\boldsymbol{r}/L$ would vary as the product $\alpha_l^2\alpha_\tau$. For the moment, however, we restrict ourselves to the condition of constant α_l and α_τ.

We next make an important assumption on the force diffusion coefficient D. We recall within the context of Fig. 7–4 and Eq. [25] that D is a function of the average lateral distance Δx, over which a force is transmitted every time it is transmitted a distance Δz. The distances Δx and Δz are very small, on the order of the size of a single particle, because they represent the "jump distance" of a force from one particle to the next. We call one pair of distances Δx and Δz an "elemental structural unit." We now ask ourselves how far laterally the force will "spread" once it has moved a vertical distance z, which involves *many* elemental structural units. The problem is equivalent to imposing a macroscopic scale change L on the z axis such that the vertical movement of the force is now $L\Delta z$, and asking how the horizontal spread distance is affected. If "on the average" our elemental structural units are always arranged in the same way relative to each other (such as if soil structure were fractal in nature), then from Fig. 7–4 we would expect the lateral spread to be $L\Delta x$. We see that the whole spreading process has been uniformly expanded by the scale factor L. Thus, for any macroscopic scale change L in a loaded granular system, the values of Δx and Δz in the definition of D in Eq. [25] are simply multiplied by L.

With this property of D, we can define *macroscopically* scaled position variables $x_M^* = x/L$, $z_M^* = z/L$ and substitute them into Eq. [24] and [25] to obtain scaled equations that are identical in form to Eq. [26] and [27]. The only difference is that the microscopically reduced variables x/α_l and z/α_l of Eq. [26] and [27] are now replaced by the macroscopically scaled x/L and z/L. The macroscopic scaling property of equations similar to Eq. [24] and [25] has already been pointed out by Hill and Harr (1982).

The implication of this result is that the distribution of interparticle forces at different relative positions $\mathbf{r}/L$ in a granular medium at constant particle size is invariant under changes in macroscopic scale, provided the boundary stress distribution is scaled accordingly as defined by Eq. [29]. Referring back to our comparison of Fig. 7–3a and 7–3c, we see then that the expected (most probable) interparticle contact forces at the macroscopically similar positions P and Q must be equal.

The previous analyses now allow solution of the third problem illustrated in Fig. 7–3b and 7–3c, where the macroscopic scale L is held constant but the microscopic scale α_l is allowed to vary. We first recall from Case 1 (L/α_l = constant) that the interparticle forces at the same relative (reduced) positions P and O in Fig. 7–3a and 7–3b are related by Eq. [21]

$$f_k = (\alpha_\tau \alpha_l^2) f_{k\tau}^*$$

But we also just established from dimensional analysis of our probabilistic equations that, for a given α_τ and α_l, that interparticle contact forces at macroscopically similar positions P and Q in Fig 7–3a and 7–3c were identical. Following simple rules of logic, because Q is equal to P and P is related to O by Eq. [21], then Q and O in Fig. 7–3b nd 7–3c must also be related by Eq. [21].

This triangular reasoning can be summarized in a simple principle. This is that, under conditions of macroscopic similar loading, the interparticle force probability at a given macroscopically scaled position $\mathbf{r}/L$ is always proportional to the scalar product $\alpha_\tau \alpha_l^2$, regardless of whether microscopic and macroscopic scale are varied independently of each other or not. This will be referred to as the macroscopic scaling principle.

The results of Braunack and Dexter (1978) in uniaxial compression tests on beds of soil aggregates appear to confirm such a macroscopic scaling effect. In these experiments, aggregate size was varied but the external dimensions of the aggregate beds and the loading system were kept constant. This was equivalent to varying microscopic scale α_l but keeping the macroscopic scale L constant. Results indicated α_l the external compressive stress required to cause fracture of a given fraction of the aggregates in a bed varied directly with the contact force f_k required to cause rupture of a single aggregate. The latter in turn was found to vary directly with α_l^2. A little thought reveals these results are consistent with the scaling principle that f_k at all positions $\mathbf{r}/L$ is proportional to the factor $\alpha_\tau \alpha_l^2$ under independent variation of microscopic relative to macroscopic scale.

At this point we can relate our findings to the continuum concept of intergranular stress. We first note that the particle surface area over which one interparticle force is acting will be proportional to α_l^2 for similar media. The interparticle force itself, however, is also proportional to α_l^2, because of the macroscopic scaling principle given by Eq. [21]. Because the intergranular stress acting on a particle surface is the ratio of the force to the surface area, it follows that the scale factor α_l^2 cancels out and thus the stress remains unaffected by a change in microscopic scale. The implication is that the intergranular stress in similarly loaded similar media is only a function of $\boldsymbol{r}/L$, independently of microscopic scale α_l or macroscopic scale L.

The fact that the reduced spatial distribution (in terms of $\boldsymbol{r}/L$) of both the interparticle forces $\boldsymbol{f}_k$ and their associated "stresses" $\boldsymbol{f}_k/\alpha_l^2$ are independent of macroscopic scale (L) indicates that the concepts of force and stress can be used interchangeably if we are only dealing with macroscopic scaling differences. Harr (1977), for example, used the concept of intergranular stress rather than force in his discussion of diffusion equations similar to our Eq. [24] and [25].

The macroscopic scaling property of internal stresses resulting from boundary loads can also be derived from continuum theory. We write the divergence theorem for the stress in a loaded body as (Malvern, 1969).

$$\int_A \boldsymbol{\tau}' \cdot \boldsymbol{n}\, dA = \int_V (\mathbf{div}\, \boldsymbol{\tau}')\, dV \qquad [30]$$

and note that, for an isotropic material, it can be reduced to the dimensionless form

$$\int_{A^*} \boldsymbol{\tau}'^* \cdot \boldsymbol{n}\, dA^* = \int_{V^*} (\mathbf{div}^*\, \boldsymbol{\tau}'^*)\, dV^* \qquad [31]$$

where $\boldsymbol{\tau}'^* = \boldsymbol{\tau}'/\alpha_\tau$, $\mathbf{div}^* = L\,\mathbf{div}$, $A^* = A/L^2$ and $V^* = V/L^3$. Here A is the loaded surface area, V is the interior volume of the material, **div** is the divergence operator, and the remaining symbols mean the same as in other parts of this chapter. The divergence theorem in the form Eq. [30] simply states the physical requirement that the net force acting on a body in the absence of gravity is equal to the vector sum of the forces acting over its surface. We see from its scaled version Eq. [31] that the reduced stress $\boldsymbol{\tau}'^*$ at any macroscopically scaled position $\boldsymbol{r}/L$ within the medium is invariant upon macroscopic changes in scale. This property underlies the use of scaled influence charts in soil mechanics to predict stresses beneath loaded surfaces (see for example Lambe & Whitman, 1969).

FORCE SYSTEM 2: EQUILIBRIUM UNDER GRAVITATIONAL FORCES

In a particulate medium where gravity is the only external force acting on the system (Fig. 7–5), the conditions of static equilibrium are given by equations similar to Eq. [2] to [5]

$$\sum_{k=1}^{K} \boldsymbol{f}_k + mg\boldsymbol{z} = 0 \qquad [32]$$

$$\sum_{k=1}^{K} (\boldsymbol{f}_k \times \boldsymbol{r}_{km}) = 0 \qquad [33]$$

$$\sum_{j=1}^{J} \boldsymbol{F}_j + Mg\boldsymbol{z} = 0 \qquad [34]$$

$$\sum_{j=1}^{J} (\boldsymbol{F}_j \times \boldsymbol{r}_{jm}) = 0 \qquad [35]$$

where m is the mass of a given particle, M is the mass of the entire system, g is the gravitational constant, z is the unit vector in the direction of the gravitational force, and all other variables are as defined before.

By noting that $m = \rho_p v$ and $M = \sum_{i=1}^{I} m_i = \rho_p V$

(where i = number of particles in the system, ρ_p is the particle density, v is the particle volume, and V is the sum of particle volumes), Eq. [32] to

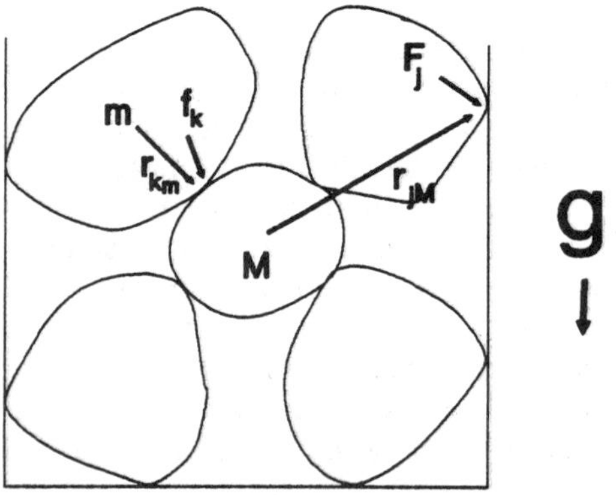

Fig. 7–5. Boundary and interparticle forces in a particle system in a gravitational field.

[35] can be reduced to

$$\sum_{k=1}^{K} \boldsymbol{f}_k^* + v^* \boldsymbol{z} = 0 \qquad [36]$$

$$\sum_{k=1}^{K} (\boldsymbol{f}_k^* \times \boldsymbol{r}_{km}^*) = 0 \qquad [37]$$

$$\sum_{j=1}^{J} \boldsymbol{F}_j^* + V^* \boldsymbol{z} = 0 \qquad [38]$$

$$\sum_{j=1}^{J} (\boldsymbol{F}_j^* \times \boldsymbol{r}_{jm}^*) = 0 \qquad [39]$$

where

$$\boldsymbol{f}_k^* = \boldsymbol{f}_k / \rho_p g \alpha_l^3 \qquad \boldsymbol{r}_{km}^* = \boldsymbol{r}_{km} / \alpha_l$$

$$\boldsymbol{F}_j^* = \boldsymbol{F}_j / \rho_p g \alpha_l^3 \qquad v^* = v / \alpha_l^3$$

$$\boldsymbol{r}_{jm}^* = \boldsymbol{r}_{jm} / \alpha_l \qquad V^* = V / \alpha_l^3$$

These results imply that the reduced forces $\boldsymbol{f}_k^*$ and $\boldsymbol{F}_j^*$ at the respective reduced positions $\boldsymbol{r}_{km}^*$ and $\boldsymbol{r}_{jm}^*$ are invariant upon changes in microscopic scale α_l. But what happens if we change the macroscopic scale L of our system? Equations [35] to [39] only describe the effects of microscopic length in systems that are exact geometrical analogues of each other, i.e., where the macroscopic scale L must vary in direct proportion to microscopic scale α_l. We would like to know the effect of changing L independently of microscopic scale.

In analyzing this problem, we note that the force acting on a given particle within a granular system in a gravitational field is the sum of two components: (i) the weight of the particle itself; and (ii) the force that is *transmitted* to the particle by its nearest neighbors due to their own weight and forces that have in turn been transmitted to them by other particles. The weight of the particle is given by $\rho_p g v \boldsymbol{z}$, where the variables involved are the same as in Eq. [32] to [39], and the transmitted force component is simply the force probability given by the recurrence relation Eq. [22]. Thus, we can write the force probability for our particle under gravity as

$$f_z(x, z + \Delta z) = 1/2[f_z(x - \Delta x, z) + f_z(x + \Delta x, z)] + \rho_p g v \boldsymbol{z} \qquad [40]$$

Performing the same manipulations we did on Eq. [22] yields the difference equation

$$\Delta f_z / \Delta z = [-1/2\, \Delta x^2 / \Delta z][\Delta(\Delta f_z / \Delta x)/\Delta x] + \rho_p g v \boldsymbol{z} / \Delta z \qquad [41]$$

which can be written in differential form as

$$\delta f_z / \delta z = D\, \delta^2 f_z / \delta x^2 + \rho_p g v \boldsymbol{z} / (\Delta z) \qquad [42]$$

where the diffusion coefficient D is as defined in Eq. [25]. We have left

(Δz) in difference rather than differential form in the denominator of the second term on the right of Eq. [42], to emphasize it is a geometrical property of our particle and not part of a differentiation operation.

Notice the similarity of Eq. [42] to the Richards Equation of water flow in soils, where movement is governed by a diffusion term and a body force term. An important difference is that Eq. [42] is linear.

To scale Eq. [42], we first reduce the variables in the body force term by multiplying the entire equation through by $1/(\rho_p g \alpha_l^2)$, noting that v is proportional to α_l^3 and (Δz) is proportional to α_l. This yields

$$(1/\rho_p g \alpha_l^2)\, \delta f_z/\delta z = (1/\rho_p g \alpha_l^2)\, D\delta^2 f_z/\delta x^2 + v z^*/(\Delta z)^* \qquad [43]$$

where $v^* = v/\alpha_l^3$ and $(\Delta z)^* = (\Delta z)/\alpha_l$. The next step is to reduce the position variables x and z of the system. Because we are interested in how the system behaves on *macroscopic* scaling of its spatial coordinates, we reduce all the x and z values of Eq. [42] (including in the diffusion coefficient D as was done for Eq. [25]) to x/L and z/L. This is achieved by multiplying Eq. [43] by the scale factor L and performing some algebraic manipulation to obtain our final reduced equation

$$\delta f_z^*/\delta z_M^* = D_M^* \delta f_z^*/\delta x_M^{*2} + v^* \mathbf{z}/(\Delta z)^* \qquad [44]$$

where the reduced variables are defined as

$$f_z^* = f_z/(\rho_p g \alpha_l^2 L) \qquad (\Delta z)^* = (\Delta z)/\alpha_l$$

$$z_M^* = z/L \qquad v^* = v/\alpha_l^3$$

$$x_M^* = x/L \qquad D_M^* = -1/2\, \Delta(x/L)^2/\Delta(z/L)$$

and z is the unit vector in the direction of the gravitational field.

Equation [44] implies that, upon macroscopic scaling, the interparticle force f_z (or more generally $\boldsymbol{f}_k$) at a given reduced position $\boldsymbol{r}/L$ will be proportional to $\rho_p g \alpha_l^2 L$. Thus, we can write our gravity-induced interparticle forces as

$$\boldsymbol{f}_k = (\rho_p g \alpha_l^2 L) \boldsymbol{f}_{kg}^*(r_M^*) \qquad [45]$$

where $\boldsymbol{f}_{kg}^*$ denotes a dimensionless "gravitational" vector, which only depends on the macroscopically reduced position $r_m^* = r/L$.

Another implication of Eq. [44] is that the interparticle forces due to gravity will remain invariant upon macroscopic scale changes provided the scale factor $\rho_p g \alpha_l^2$L remains constant. Because particle density and gravity are usually relatively constant, we are left with the requirement that *the macroscopic scale length and the square of the microscopic length must vary inversely with each other* if macroscopic scaling invariance of interparticle forces is to be maintained.

Equation [45] contains the results of Eq. [36] to [39] as a special case. The latter equations considered the situation where macroscopic length L is varied in the same proportion as microscopic length α_l. In this case we can substitute α_l for L in Eq. [45], which yields

$$\boldsymbol{f}_k = (\alpha_l^3 \rho_p g)\, \boldsymbol{f}_{kg}^*(\boldsymbol{r}/\alpha_l) \qquad [46]$$

The implication of this equation is that for systems that are exact geometrical analogues of each other, the interparticle force at any microscopically reduced position $\boldsymbol{r}/\alpha_l$ will be proportional to the scalar product $\alpha_l^3 \rho_p g$, in agreement with Eq. [36] to [39].

To compare these findings to available experimental results, we modify Eq. [46] slightly by relaxing some of its requirements. We first note that when we are far enough away from the system walls that the latter have minimal influence, we can treat a particle system as a semiinfinite medium in which the only boundary is the top of the soil profile. If, furthermore, the radius of curvature of the soil surface is large relative to the depth h within the soil medium, gravitational effects can be assumed to be the same at any given depth h. For situations where such an assumption holds, the reduced interparticle forces $\boldsymbol{f}_k^*$ of Eq. [46] can be considered functions of a new reduced "position" variable $h^* = h/\alpha_l$. Experiments by Whitely and Dexter (1982) on the forces required to move single glass spheres through beds of similar-sized spheres have shown excellent agreement with this prediction.

As we did when discussing the effects of applied boundary stresses, we wish to relate our findings on interparticle forces under gravity to the continuum concept of stress. Recalling from our discussion on applied stress systems that the intergranular stress $\boldsymbol{\tau}$ is proportional to $\boldsymbol{f}_k/\alpha_l^2$, we can substitute the identity $\boldsymbol{\tau} = \boldsymbol{f}_k/\alpha_l^2$ into Eq. [45] and obtain the result that the dimensionless stress $\boldsymbol{\tau}/(\rho_p gL)$ in a gravitational field is only dependent on $\boldsymbol{r}/L$ and is thus invariant upon macroscopic scaling. A corollary of this result is that the intergranular stress $\boldsymbol{\tau}$ induced by gravity will remain invariant upon macroscopic scaling as long as the product $\rho_p gL$ is constant. Note that here, as in the case of the applied stress system, the stress is independent of microscopic scale.

The same result can be obtained from purely continuum considerations. We write the divergence theorem for a material element of density ρ_p and volume V at equilibrium under gravity as (Malvern, 1969)

$$\int_V (\mathbf{div}\ \boldsymbol{\tau})\, dV = \int_V (\rho_p g \boldsymbol{z})\, dV \qquad [47]$$

where g is the gravitational constant and $\boldsymbol{z}$ is the unit vector in the direction of the gravitational field. Equation [47] simply states that at equilibrium the gravitational force on the element must be balanced by the stresses required to keep it from accelerating. Dividing through by $\rho_p g$, applying the macroscopic scaling identities **div*** = L **div** and $V^* = V/L^3$, and using the linear property (b) (**div** $\boldsymbol{\tau}$) = **div** $(b\boldsymbol{\tau})$, where b is any constant, yields

$$\int_{V^*} (\mathrm{div}^* \tau^*)\, dV^* = \int_{V^*} z dV^* \qquad [48]$$

where $\boldsymbol{\tau}^* = \boldsymbol{\tau}/L\rho_p g$. This result tells us two important things. The first is that the quantity $\boldsymbol{\tau}/L\rho_p g$ will remain invariant upon macroscopic scale changes. Thus, if the gravity-induced stresses $\boldsymbol{\tau}$ are to remain invariant under scaling, the scalar product $L\rho_p g$ must also remain constant. In practical terms, this means we would need to change L and the body force $\rho_p g$ inversely to each other in any scaling experiment where we wanted τ to be only a function of the scaled position $\boldsymbol{r}/L$. The second conclusion from Eq. [48] is that for any given $\boldsymbol{r}/L$ under changes in scale, the magnitude of $\boldsymbol{\tau}$ will always be proportional to $L\rho_p g$. Note both conclusions are in agreement with those derived earlier parting from a microscopic viewpoint.

FORCE SYSTEM 3: EQUILIBRIUM UNDER PORE LIQUID STRESSES

The condition of static equilibrium for a particle framework under the action of pore liquid phase (Fig. 7–6) requires that the sums of all forces

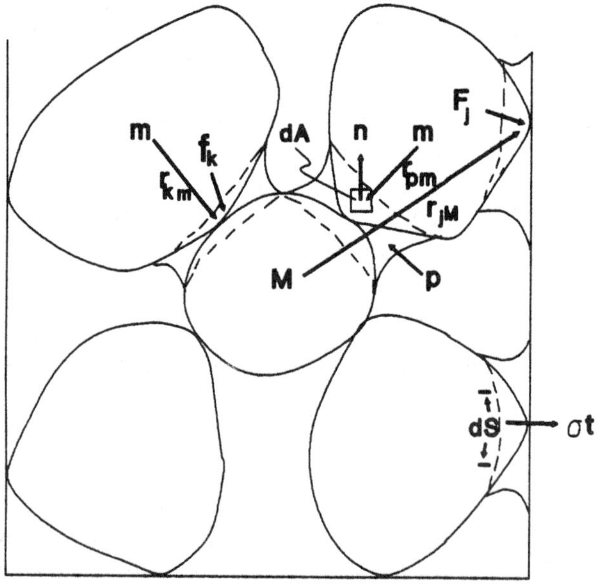

Fig. 7–6. Boundary and interparticle forces in a granular medium induced by the liquid phase in the presence of a gas-liquid interface.

and torques induced by the liquid must be zero. The equations of equilibrium for each individual particle are

$$\sum_{k=1}^{K} \boldsymbol{f}_k + \int_{A_p} p\mathbf{1} \cdot \boldsymbol{n}\, dA + \int_{S_p} \sigma \boldsymbol{t}\, dS = 0 \qquad [49]$$

and

$$\sum_{k=1}^{K} (\boldsymbol{f}_k \times \boldsymbol{r}_{km}) + \int_{A_p} [(p\mathbf{1} \cdot \boldsymbol{n}) \times \boldsymbol{r}_{pm}]\, dA + \int_{S_p} (\sigma \boldsymbol{t} \times \boldsymbol{r}_{tm})\, dS \qquad [50]$$

Here p is the capillary pressure deficiency, $\mathbf{1}$ is the unit tensor, $\boldsymbol{r}_{pm}$ is a vector indicating the position of a wetted particle surface element dA relative to the center of mass m of the particle, and n is the unit vector normal to dA. The total wetted area of the particle is denoted by A_p. The parameter σ is the surface tension of the liquid-gas interface, $\boldsymbol{t}$ is a unit vector perpendicular to the liquid-gas-solid interface (the dashed lines in Fig. 7–6), which indicates the direction of action of the surface tension force, and $\boldsymbol{r}_{pm}$ is a vector indicating the position of a given segment dS of the liquid-gas-solid interface relative to the particle center of mass. The total length of the liquid-gas-solid interface for the particle is S_p. The vectors $\boldsymbol{f}_k$ and $\boldsymbol{r}_{km}$ have their usual meaning of the interparticle force and its position relative to the particle center of mass.

Similar equations describe the liquid phase effects on the boundary of the system

$$\sum_{j=1}^{J} \left[\boldsymbol{F}_j + \int_{A_{bj}} p\mathbf{1} \cdot \boldsymbol{n}\, dA + \int_{S_{bj}} \sigma \boldsymbol{t}\, dS \right] = 0 \qquad [51]$$

and

$$\sum_{j=1}^{J} \left[(\boldsymbol{F}_j \times \boldsymbol{r}_{jM}) + \int_{A_{bj}} [(p\mathbf{1} \cdot \boldsymbol{n}) \times \boldsymbol{r}_{bM}]\, dA + \int_{S_{bj}} (\sigma \boldsymbol{t} \times \boldsymbol{r}_{tM})\, dS \right] \qquad [52]$$

The variables p, σ, $\mathbf{1}$, and $\boldsymbol{t}$ have the same meaning as in Eq. [49] and [50]. The remaining variables in Eq. [51] and [52] are analogous to their counterparts in Eq. [49] nd [50]. The position vectors $\boldsymbol{r}_{jM}$, and $\boldsymbol{r}_{bM}$, and $\boldsymbol{r}_{tM}$, respectively, relate surface contact forces $\boldsymbol{F}_j$, wetted surface elements dA, and liquid-gas-solid interface segments dS on the boundary to the center of mass of the system. The variables A_{bj} and S_{bj} are the wetted area and length of the liquid-gas-solid interface on the boundary, respectively, due to the liquid meniscus associated with the jth particle along the boundary.

The liquid phase in the system is not required to be in static equilibrium, provided the condition of "creeping flow" is maintained such that inertial forces associated with the flow can be neglected. Liquid pressure gradients, arising from spatial gradients in the curvature of the gas-liquid interface, will not cause a net force and hence an acceleration on the system as a whole; they will simply cause particles in some areas to be pulled against each other with greater force that in other areas. No acceleration will result, because particles that are being pulled against *each other* have nowhere to move. The same argument applies to the boundary of the system: it will simply be held tighter against the surface particles at some points that at others. For practical purposes the boundary can simply be considered as another particle that is stuck on to the system.

Our next step is to analyze the scaling properties of the system, in terms of their effects on the spatial distribution of interparticle forces. We wish to establish the effects of both microscopic and macroscopic changes in scale.

Our independent variable in the system is the distribution of the liquid pressure deficiency p and the associated microscopic geometry of the liquid phase. The behavior of pore liquids in soils is rather complex, but fortunately the scaling properties of Newtonian fluid dynamics in unsaturated granular media have been worked out by Miller and Miller (1956). Here we summarize the results of their theory, which are of use to us.

The behavior of a Newtonian fluid in granular similar media can be summarized by the invariant form of the Richards Equation (Miller, 1980)

$$\delta\Theta^*(p^*)/\delta t^* = \mathbf{div}^* \, [\boldsymbol{K}^*(p^*) \cdot (\boldsymbol{f}_l^* - \overline{\boldsymbol{V}}{}^* p^*)] \qquad [53]$$

where the dimensionless variables are defined as

$$\Theta^* = \text{volumetric pore liquid content}$$

$$p^* = [(\alpha_l/\sigma)p]$$

$$t^* = [(\sigma/\eta)(\alpha_l/L^2)t]$$

$$\boldsymbol{f}_l^* = [(L\alpha_l\rho_l/\sigma)g\boldsymbol{z}]$$

$$\mathbf{div}^* = [L \, \mathbf{div}]$$

$$\boldsymbol{K}^* = [(\eta/\alpha_l^2)\boldsymbol{K}]$$

Here t is time, η and ρ_l are the respective kinematic viscosity and density of the liquid, σ is surface tension, p is the liquid pressure deficiency, $\boldsymbol{K}$ is the hydraulic conductivity tensor, **div** represents the divergence operator, $\overline{\boldsymbol{V}}$ is the gradient operator, g is the gravitational constant, $\boldsymbol{z}$ is the unit vector in the direction of the gravitational field, α_l is microscopic characteristic length, and L is the macroscopic characteristic length.

The implication of Eq. [53] is that, for a given reduced liquid body force $\boldsymbol{f}_l^*$ and reduced initial conditions $[\Theta^*, p^*](\boldsymbol{r}_M^*)$, the values of Θ^* and

p^* at any *macroscopically* reduced position $\boldsymbol{r}_M^* = \boldsymbol{r}/L$, and time t^* in granular similar media will always be constant. Because the body force $\boldsymbol{f}_l^*$ must remain constant, and the variables g, ρ_l and σ are in most cases relatively constant, it follows from the definition of $\boldsymbol{f}_l^*$ in Eq. [53] that the microscopic α_l and the macroscopic L must usually vary inversely to each other for similitude to be maintained.

With this result in mind, we now return our attention to the particle equilibrium systems in Eq. [49] to [53]. Our first consideration regards the scaling properties of the equations for single particles given by Eq. [49] and [50]. Because all the spatial dimensions involved are on a particle-size scale, we can only use the microscopic α_l as our length scaling factor. The fluid pressure deficiency p is scaled as defined for Eq. [53]. With these relations we can write Eq. [49] and [50] in reduced form as

$$\sum_{k=1}^{K} \boldsymbol{f}_k^* + \int_{A_p^*} p^* \mathbf{1} \cdot \boldsymbol{n}\, dA^* + \int_{S_p^*} \boldsymbol{t}\, dS^* = 0 \qquad [54]$$

and

$$\sum_{k=1}^{K} (\boldsymbol{f}_k^* \times \boldsymbol{r}_{km}^*) + \int_{A_p^*} [(p^* \mathbf{1} \cdot \boldsymbol{n}) \times \boldsymbol{r}_{pm}^*]\, dA^* + \int_{S_p^*} (\boldsymbol{t} \times \boldsymbol{r}_{tm}^*)\, dS^* \qquad [55]$$

where

$$\boldsymbol{f}_k^* = (\boldsymbol{f}_k/\sigma\alpha_l)$$

$$\boldsymbol{r}_{km}^* = (\boldsymbol{r}_{km}/\alpha_l)$$

$$\boldsymbol{r}_{pm}^* = (\boldsymbol{r}_{pm}/\alpha_l)$$

$$\boldsymbol{r}_{tm}^* = (\boldsymbol{r}_{tm}/\alpha_l)$$

$$p^* = [(\alpha_l/\sigma)p]$$

$$A^* = (A/\alpha_l^2)$$

$$A_p^* = (A_p/\alpha_l^2)$$

$$S^* = (S/\alpha_l)$$

$$S_p^* = (S_p/\alpha_l)$$

We note immediately that, because of the definition of $\boldsymbol{f}_k^*$, which results from scaling Eq. [49] and [50], the interparticle forces $\boldsymbol{f}_k$ are proportional to the scalar product ($\sigma\ \alpha_l$). This is useful information, but how can it be linked to the *macroscopic* position vectors of the reduced $\boldsymbol{f}_k^*$ values? These vectors do not appear explicitly in Eq. [54] and [55]. However, they are implied by the presence of p^*, the position of which, according to Eq. [53], must scale as $\boldsymbol{r}/L$. Because a given value of $\boldsymbol{f}_k^*$ is determined by the specific values of p^* and the associated geometrical parameters A_p^* and S_p^* that occur in the immediate neighborhood it follows the position of $\boldsymbol{f}_k^*$ must also scale as $\boldsymbol{r}/L$.

We now turn to the scaling properties of the boundary conditions Eq. [51] and [52]. We first note the position vectors $\boldsymbol{r}_{jM}$, $\boldsymbol{r}_{bM}$, and $\boldsymbol{r}_{tM}$ relate the boundary forces to the center of mass of the entire system. Hence, they are *macroscopic* parameters and accordingly must scale as $\mathbf{1}/L$. The parameters dA, dS, A_{bj}, S_{bj}, and P, on the other hand, are *microscopic* in scale because they are associated with the size and shape of a contact zone between a single particle and the boundary. Scaling all variables accordingly reduces Eq. [51] and [52] to the invariant forms

$$\sum_{j=1}^{J} [f_j^* + \int_{A_{bj}^*} p^*\mathbf{1} \cdot \boldsymbol{n}\, dA^* + \int_{S_{bj}} \mathbf{t}\, dS^*] = 0 \qquad [56]$$

and

$$\sum_{j=1}^{J} \{(\boldsymbol{F}_j^* \times \boldsymbol{r}_{jM}^*) + \int_{A_{bj}^*} [(p^*\mathbf{1} \cdot \boldsymbol{n}) \times \boldsymbol{r}_{bM}^*]\, dA^* + \int_{S_{bj}} (\boldsymbol{t} \times \boldsymbol{r}_{tM}^*)\, dS^*\} = 0 \qquad [57]$$

where

$$\boldsymbol{F}_j^* = (\boldsymbol{F}_j/\sigma\alpha_l)$$

$$\boldsymbol{r}_{jM}^* = (\boldsymbol{r}_{jM}/L)$$

$$\boldsymbol{r}_{bM}^* = (\boldsymbol{r}_{bM}/L)$$

$$\boldsymbol{r}_{tM}^* = (\boldsymbol{r}_{tM}/L)$$

$$p^* = [(\alpha_l/\sigma)p]$$

$$A^* = (A/\alpha_l^2)$$

$$A_{bj}^* = (A_{bj}/\alpha_l^2)$$

$$S^* = (S/\alpha_l)$$

$$S_{bj}^* = (S_{bj}/\alpha_l)$$

Equations [56] and [57] together with Eq. [53] imply the same scaling for boundary forces $\boldsymbol{F}_j^*$ as were found for interior forces $\boldsymbol{f}_k^*$ in Eq. [54] and [55]. This is of course a requirement if our problem is formulated correctly. We note the definitions of $\boldsymbol{F}_j^*$ and $\boldsymbol{f}_k^*$, which result from scaling, imply both $\boldsymbol{F}_j$ and $\boldsymbol{f}_k$ at a given $(\boldsymbol{r}_m^*, t^*)$ vary as the product $\sigma\ \alpha_l$. This allows us to write f_k (or $\boldsymbol{F}_j$ if we chose to) as

$$\boldsymbol{f}_k = (\sigma\alpha_l)\boldsymbol{f}_{kl}^*(\boldsymbol{r}_M^*, t^*) \qquad [58]$$

subject to invariance of $\boldsymbol{K}^*(\Theta^*), \boldsymbol{f}_l^*$ and the initial conditions $\Theta^*(p^*)$ for the system. The parameter $\boldsymbol{f}_{kl}^*(\boldsymbol{r}_M^*, t^*)$ denotes a dimensionless vector that varies only with $\boldsymbol{r}_m^*$ and t^*. Note that t^* does not enter the problem for the special case when the liquid phase is in equilibrium. Similarly to what we found when analyzing gravitational and applied stress effects on our system, the interparticle contact forces in Eq. [58] can always be resolved into a scalar component and a dimensionless vector that is independent of scale.

It is again illustrative to compare the results of our microscopic analysis with continuum concepts that have been applied to soils. If we divide Eq. [58] by α_l^2 to convert f_k to a "stress" $\boldsymbol{\tau}$ (actually a vector with units of stress), and introduce the identity $p^* = p(\alpha_l/\sigma)$, we obtain

$$\boldsymbol{f}_k/\alpha_l^2 = \boldsymbol{\tau} = p[(\mathbf{1}/p^*)\boldsymbol{f}_{kl}^*(\boldsymbol{r}_M^*, t^*)] \qquad [59]$$

Being essentially a dimensionless geometrical property of an unsaturated soil, the parameter $\mathbf{1}/p^*\boldsymbol{f}_{kl}^*(\boldsymbol{r}_M^*, t^*)$ can be regarded as a function (which we denominate by the Greek letter $\boldsymbol{\chi}$) of the water content Θ^*. Substituting this into Eq. [59] gives

$$\boldsymbol{\tau} = p\boldsymbol{\chi}[\Theta^*(\boldsymbol{r}_M^*, t^*)] \qquad [60]$$

This expression is similar to the term $pX(^*)$, which has been used to denote pore-liquid effects in effective stress equations (Aitchison, 1961). Notice, however, that $p\boldsymbol{\chi}(^*)$ in Eq. [60] is a *vector* that possesses not only magnitude but also a direction. This direction is not necessarily equal to the direction of intergranular forces of equal magnitude caused by boundary stresses (Burland, 1965). Directional differences between the effects of applied stresses and fluid pressures on intergranular forces are usually not taken into account in effective stress equations, and can represent a serious drawback to using the latter to characterize effective stresses (Burland, 1965; Matyas & Radhakrishna, 1968; Fredlund & Morgenstern, 1977). More will be said about directional effects in the following discussion.

SUPERPOSITION OF FORCE SYSTEMS

Having derived expressions for the interparticle forces in each of our three equilibrium systems, we now superimpose them to obtain a final solution. Addition of the component solutions Eq. [21], [45], and [58] yields

$$\boldsymbol{f}_{ks} = (\alpha_\tau \alpha_l^2)\boldsymbol{f}^*_{kT}(\boldsymbol{r}^*_M) + (\rho_p g \alpha_l^2 L)\boldsymbol{f}^*_{kg}(\boldsymbol{r}^*_M) + (\alpha_l \sigma)\boldsymbol{f}^*_{kl}(\boldsymbol{r}^*_M, t^*) \quad [61]$$

where $\boldsymbol{f}_{ks}$ denotes the resultant interparticle forces obtained after superposition.

IMPLICATIONS—GENERAL ASPECTS

We need to take a good look at the implications of what we have obtained.

Equation [61] gives us a rule for the functional relationships between the interparticle forces $\boldsymbol{f}_{ks}$ and different scalar parameters in a loaded unsaturated soil under gravity, such that these relationships will always be invariant upon time and macroscopic position scaling.

We have a complication, however, in that each of the three terms on the right of Eq. [61] has a different constraint under which it will allow interparticle forces to remain invariant upon scaling. The first (applied stress) term has no constraint other than similar loading. The second term, however, which gives the effect of gravity, has the constraint that the product $(\rho_p g \alpha_l^2 L)$ must remain constant as we saw in analyzing Eq. [45]. The liquid phase term, because of the scaling relation Eq. [53], requires that $(\rho_l \alpha_l L/\sigma)g$ be constant. Under what conditions will all three terms be invariant on scaling? We find this "common ground" by substituting any one of the gravitational or liquid phase constraints into the other and obtaining the additional constraint

$$(\rho_p \alpha_l \sigma / \rho_l) = \text{constant} \quad [62]$$

What this means is that whenever *either* the liquid phase or the gravitational constraint *and* Eq. [62] are satisfied, the third constraint will also be satisfied. Any solution of Eq. [62] that meets these requirements will scale as $\boldsymbol{r}/L$.

To give an idea of how this works, suppose we choose to satisfy the liquid phase constraint by varying α_l and L inversely to each other. This is equivalent to writing $\alpha_l = k/L$, where k is a proportionality constant. Substituting this into our compatibility condition, Eq. [62] yields

$$(\rho_p \sigma)/(\rho_l L) = \text{constant}$$

Such a condition will be satisfied if, for example, we require that L and σ vary directly with each other. Thus, we have that all solutions of Eq. [61] will scale with r/L as long as both $L\alpha_l$ and σ/L are kept constant. We could play similar games with the other variables in the constraint equations.

Suppose, however, we encounter a situation where no two constraints are satisfied but we want to try to "scale" our problem anyway using only one constraint. How could we make an informed guess on whether this omission will cause serious distortion of reality in our scale models? As an aid in such a decision we convert all variables of Eq. [61] to "stresses" by dividing through by α_l^2 and substituting the identity $\sigma/\alpha_l = p/p^*$ of Eq. [53] to obtain

$$f_{ks}/\alpha_l^2 = \boldsymbol{\tau} = (\alpha_\tau)f_{kT}^*(\boldsymbol{r}_M^*) + (\rho_p gL)f_{kg}^*(\boldsymbol{r}_M^*) + (p/p^*)f_{kl}^*(\boldsymbol{r}_M^*, t^*) \quad [63]$$

where now all our scalar coefficients on the right have dimensions of stress and furthermore can be measured or at least roughly estimated. Note that α_τ is the "magnitude" of the applied stress, p can be estimated tensiometrically and the "gravitational stress" can be approximated as $\rho_p gL$ by taking L as maximum depth within the soil profile. For example, suppose the bulk density of a soil were 1.2 Mg/m^3, the "average" liquid pressure deficiency were 0.1 MPa and the "average" applied stress were 0.2 MPa. We wish to vary our scale length between 0 and 0.2 m. How much error would be involved in ignoring gravity? A quick calculation shows that at the maximum depth of 0.2 m, the gravitational stresses would be on the order of 0.0024 MPa, which is quite low compared with the other stresses. We could thus probably scale our problem using only the liquid phase constraint $[(\rho_l\alpha_l L)/\sigma]g$ = constant, as long as α_l (pore size) is not allowed to increase to the point where pore-water pressures (p) are reduced to the same order of magnitude as the gravitational stress $\rho_p gL$.

It is important to understand the meaning and practical implications of the dimensionless vector functions $f_{kT}^*(\boldsymbol{r}_M^*)$, $f_{kg}^*(\boldsymbol{r}_M^*)$, and $f_{kl}^*(\boldsymbol{r}_M^*, t^*)$ in Eq. [61] and [63]. They represent the spatial-temporal distribution of the directions and *relative* magnitudes of interparticle forces in a granular medium, in relation to the respective geometries of the particle framework and applied force systems. They will change accordingly as the geometry of the system is changed. The latter will vary with any change in the similar loading criterion $\overline{V}_M^*\boldsymbol{\tau}'^*$ or parameters such as void ratio and water content, which are requirements of Miller and Miller similarity. Note that void ratio is very important to soil structural strength because of its relation to the proximity of particles to each other. In scaling studies involving only liquid phase behavior, it is often found that using the degree of pore saturation gives better results than using volumetric liquid content (Reichart et al., 1972; Warrick et al. 1977). This gets around having to deal with differences in void ratio. But why does it work? What is tacitly assumed is that the "effective porosity," where most of the liquid phase activity takes place, is always similar in shape regardless of void ratio. As long as the "effective porosity" remains similar, for practical purposes one can forget about whether the solid phase is similar or not. In dealing with the strength properties of the particle matrix, however, the similarity of *both* solid and liquid phases is important. Thus, void ratio, along with degree of saturation and liquid pressure deficiency, is an extremely important macroscopic soil parameter, which needs to be taken into account in explaining the variability of soil mechanical properties. These measurements should also be accompanied by microscopic indices of similarity such as particle-size distribution and particle shape.

The directional aspect of the dimensionless vectors in Eq. [61] and [63] is very important, because the directions of action of the force components that the vectors represent are not always the same. Burland (1965) has pointed out that when a stress is applied to the boundary of a granular system, even if macroscopically the applied stress is entirely normal to the surface (as would occur under hydrostatic loading), the tendency of the particles to roll and slide over each other in response to the stress will result in slippage (shear) as well as normal forces at the interparticle points of contact. On the other hand, the interparticle forces induced by the liquid meniscii, which surround the particle contact points, are primarily *normal* to the particle surfaces. Gravitational forces can be expected to have still different directional effects. Thus, the net intergranular force at a point must treated as the vector sum of the individual effects, as indicated in Eq. [61] and [63]. Burland (1965), Matyas and Radhakrishna (1968), and others have emphasized the limitations of many effective stress equations that tacitly assume normal applied stresses and liquid phase effects act in the same direction.

IMPLICATIONS FOR SOIL FAILURE CRITERIA

Our analysis until now has considered changes in the magnitude of intergranular forces and stresses in similarly loaded similar media, as a function of changes in the scalar system parameters L, α_l, α_τ, σ, g, ρ_l, ρ_p, and η. However, nothing has been said about the conditions under which these forces will cause irreversible soil deformation (failure). A few observations on this respect follow.

For the case of unsaturated soils under tension, a failure criterion that has been proposed is that intergranular stresses at crack edges must be zero at some location in the soil medium (Snyder & Miller, 1985). This corresponds to stating that tensile failure of similar media will occur whenever the left-hand side of Eq. [61] or [63] is zero. Such a criterion was implicit in the early Haines-Fisher theory cited at the beginning of this chapter. Snyder and Miller (1985) partially verified it for the special case where gravitational forces could be ignored.

Shear failure (interparticle slippage) has been postulated to occur when the ratio of tangential to normal interparticle force components reaches a value equal to the friction coefficient (μ) of the solid material. Thus, for a given configuration of the soil particle framework and constant μ, the failure condition will depend on the *geometry* (relative magnitudes and directions) of the intergranular force vector components (Dietrich, 1977). Changes in the scalar magnitude of these components should have relatively little effect on the failure condition, *provided all changes in magnitude occur in the same proportion and the friction coefficient μ remains constant.* This would seem to imply that, for similarly loaded similar media of constant μ in a similar state of incipient shear failure, the scalar groups $\alpha_\tau \alpha_l^2$, $\rho_p g \alpha_l^2 L$, and $\alpha_l \rho \sigma$ of Eq. [61] should always occur in the same ratios relative to each other.

PERSPECTIVE

Based on available knowledge, we have attempted in this chapter to develop some insight on the effects of scale changes in the parameters that influence soil mechanical behavior. Special cases of the theory have been shown to be consistent with experiments reported in the literature. However, a systematic evaluation of the theory based on mechanical testing of granular media that are similar in terms of both pore and particle geometry is necessary for its general validation.

Soil mechanical theories based on probability concepts have proven very useful in our scaling analysis. Although these theories have traditionally received little attention in soil mechanics, their fundamental relation to microscopic structural geometry could make them useful tools in soil physics for studying the influence of soil structure on its mechanical behavior.

Soil dynamics is another area of research, which, although much more complex than the static problem considered in this chapter, could perhaps profit greatly from Miller and Miller scaling theory in conjunction with statistical mechanical methods. An important application could be the study of viscous reaction of soil pore fluids to relative particle movement during deformation of unsaturated soils.

The results of our scaling analysis suggest a minimum set of soil parameters, which should be measured when attempting to explain differences in soil mechanical behavior. A data set should include the macroscopic parameters void ratio, water content and matric potential, and microscopic criteria such as the size distribution, shape, and density of soil particles.

The scaling relations and associated implications that we have developed are, strictly speaking, limited to the case of granular noncolloidal soils. The extent to which they could be extrapolated to soils with different degrees of colloidal activity remains to be seen.

DEDICATION

This chapter is dedicated to my friend and professor Dr. Robert D. Miller, whose encouragement and extraordinary physical insight were influential in development of the material presented herein.

REFERENCES

Aitchison, G.D. 1961. Relationships of moisture and effective stress functions in unsaturated soils. p. 47–52. *In* Pore pressure and suction in soils. Butterworths, London.

Braunack, M.V., and A.R. Dexter. 1978. Compaction of aggregate beds. p. 119–126. *In* W.W. Emerson et al. (ed.) Modification of soil structure. John Wiley & Sons, New York.

Brown, R.L., and J.C. Richards. 1970. Principles of powder mechanics. Pergamon, Oxford.

Burland, J.B. 1965. Some aspects of the mechanical behavior of partly saturated soils. p. 270–278. *In* G.D. Aitchison, (ed.) Moisture equilibria and moisture changes in soils beneath covered areas. Butterworths, London.

David, F.W., and H. Nolle, 1982. Experimental modelling in engineering. Butterworths, London.

Dietrich, Th. 1978. A comprehensive mechanical model of sand at low stress levels. p. 33–43. *In* S. Murayama and A.N. Schofield (ed.) Constitutive equations of soils. Japanese Soc. of Soil Mechanics and Foundation Eng., Tokyo.

Fisher, R.A. 1926. On the capillary forces in an ideal soil; correction of formulae given by W.B. Haines. J. Agric. Sci. 16:492–505.

Fredlund, D.G., and N.R. Morgenstern. 1977. Stress state variables for unsaturated soils. J. Geotech. Eng. Div., Am. Soc. Civ. Eng., 107:447–466.

Freitag, D.R., R.L. Schafer, and R.D. Wismer. 1977. Similitude studies of soil-machine systems. p. 8–19. *In* Similitude of soil-machine Systems. ASAE Publ. 3–77. Am. Soc. Agric. Eng., St. Joseph, MI.

Haines, W.B. 1925. Studies on the physical properties of soils. II. A note on the cohesion developed by capillary forces in an ideal soil. J. Agric. Sci. 15:529–535.

Harr, M.E. 1977. Mechanics of particulate media, a probabilistic approach. McGraw-Hill, New York.

Hettiaratchi, D.R.P., and J.R. O'Callaghan. 1984. Mechanical behavior of agricultural soils. J. Agric. Eng. Res. 25:239–259.

Hill, J.M., and M.E. Harr. 1982. Elastic and particulate media. J. Eng. Mech. Div. Am. Soc. Civ. Eng. 108:596–604.

Hiramatsu, Y., and Y. Oka. 1966. Determination of the tensile strength of rock by a compression test of an irregular test piece. Int. J. Rock Mech. Min. Sci. 3:89–99.

Lambe, T.W., and R.V. Whitman. 1969. Soil mechanics. John Wiley & Sons, New York.

Malvern, L.E. 1969. Introduction to the mechanics of a continuous medium. Prentice-Hall, Englewood Cliffs, N.J.

Matyas, E.L., and H.S. Radhakrishna. 1968. Volume change characteristics of partially saturated soils. Geotechnique 18:432–448.

Meriam, J.R. 1975. Statics. John Wiley & Sons, New York.

Miller, E.E. 1980. Similitude and scaling of soil-water phenomena p. 300–318. *In* D. Hillel, (ed.) Applications of soil physics. Academic Press, New York.

Miller, E.E., and R.D. Miller. 1956. Physical theory for capillary flow phenomena. J. Appl. Phys. 27:324–332.

Reichardt, K., D.R. Nielsen, and J.W. Biggar. 1972. Scaling of horizontal infiltration into homogeneous soils. Soil Sci. Soc. Am. Proc. 36:241–245.

Sergeev, I.T. 1969. The application of probability-process equations to the theory of stress distribution in non-cohesive soil foundation beds. (Translated from Russian.) Fundamenty i Mekhanika Gruntov 2:5–7.

Skempton, A.W. 1960. Significance of Terzaghi's concept of effective stress. p. 42–53. *In* From theory to practice in soil mechanics. John Wiley & Sons, New York.

Smoltczyk, H.U. 1967. Stress computation in soil media. J. Soil Mech. Found. Div. Am. Soc. Civ. Eng. 93:101–124.

Snyder, V.A., and R.D. Miller. 1985. Tensile strength of unsaturated soils. Soil Sci. Soc. Am. J. 49:58–65.

Symon, K.R. 1971. Mechanics. Addison-Wesley, Reading, MA.

Timoshenko, S.P., and J.N. Goodier. 1970. Theory of elasticity. McGraw-Hill, New York.

Towner, G.D. 1972. The assessment of soil texture from soil strength measurements. J. Soil Sci. 25:298–306.

Vomocil, J.A., and L.J. Waldron. 1962. Effect of moisture content on tensile strength of unsaturated glass bead systems. Soil Sci. Soc. Am. Proc. 26:409–412.

Warrick, A.W., G.C. Mullen, and D.R. Nielsen. 1977. Scaling field measured soil hydraulic properties using a similar media concept. Water Resour. Res. 13:325–362.

Whitely, G.M., and A.R. Dexter. 1982. Forces required to displace individual particles within beds of similar particles. J. Agric. Eng. Res. 27:215–225.

Wu, T.H. 1976. Soil mechanics. Allyn & Bacon, Newton, MA.

8 The Consequences of Fractal Scaling in Heterogeneous Soils and Porous Media

Scott W. Tyler and Stephen W. Wheatcraft

Desert Research Institute
Reno, Nevada

In light of the subject of this special publication, it is of interest to review traditional concepts of scaling. Miller (1980) provides the following definition: "*Scaling simplifies problems by expressing them in the smallest number of reduced variables. Any solution for a problem that has been worked out in reduced formulation holds true for an infinite variety of actual systems, which, although they differ physically, are simple 'scale models' of each other.*" A more general definition is provided by Shouse et al. (1989): "*Scaling is a systematic method for specifying a change of variable that transforms one system to another one with more desirable traits.*" Scaling works quite well for a large class of problems in which the porous media is homogeneous on a large scale. More generally, we can apply scaling to systems in which there is a separation of scales. This concept is illustrated in Fig. 8–1. The porous medium is heterogeneous at a small scale (Fig. 8–1a), but when we view this same material at a larger scale, it appears homogeneous (Fig. 8–1b). The reason that scaling works so well in Fig. 8–1b is that our averaging window is large compared with the scale of heterogeneity.

The relatively recent concepts of fractal mathematics have provided us with the opportunity to develop new conceptual models for the way in which we view heterogeneity. Contrast Fig. 8–2 with Fig. 8–1. In Fig. 8–2, we see three major lenses, or pockets of heterogeneity, but inside each lens, there are three more lenses. This process continues on at least several smaller "recursive" levels, and conversely, there may be larger lenses surrounding the three major lenses. This model of heterogeneity is referred to as a *self-similar process*. One of its primary characteristics is that it looks the same at any scale (or at least over a wide range of scales in the real world), a property that we refer to as *fractal scaling*. A primary consequence

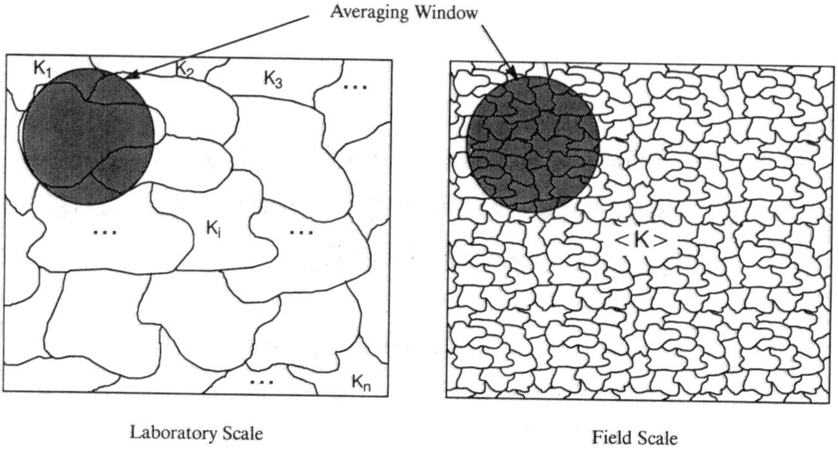

Fig. 8–1. Separation of scales that leads to traditional scaling theory.

of fractal scaling is that traditional scaling techniques cannot be applied to materials that exhibit fractal scaling.

Fractal mathematics and fractal scaling have been applied to a wide range of physical systems. The applications have ranged from simple description, i.e., a more physically realistic description of such things as topographic relief to the development of physically consistent theories on the formation and nature of fluid turbulence. Because many of the shapes and processes directly impact (or are strongly impacted by) the formation of soils and other geologic formation, it is a logical course of research to

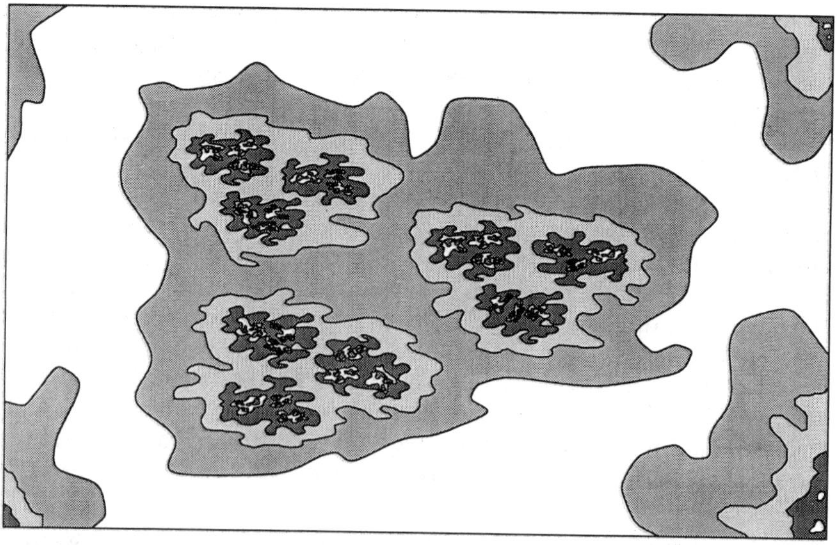

Fig. 8–2. Self-similar (fractal) model of heterogeneity.

apply the tools of fractal mathematics to the variability of hydraulic properties of field soils and aquifers.

In this chapter we will begin with a brief overview of fractal scaling and its basic results. This will be followed by a qualitative of analysis of the nature and form of heterogeneity in field soils. We will close the chapter with a simple example of fractal scaling as applied to soil water retention.

FRACTAL SCALING CONCEPTS

In simple terms, fractal scaling may be quantified as *the closer you look, the more you see*. Fractal mathematics provide the rules that determine *what you see*. The concepts of fractals are based on the notion of scale-invariant transforms; transformation in which the object or process is mapped onto itself in a geometrically similar manner.

The most simple example of such a mapping is given by the Koch curve (Mandelbrot, 1983; Feder, 1988). The initiator (Feder, 1988) of the Koch curve is given in Fig. 8–3a. with fixed points at (0,0), (1/3,0), (1/2,1/3), (2/3,0), and (1,0). Between each of the initial points, a scale-invariant transform of the form

$$(x_{i+1}, y_{i+1}) = f(x_i, y_i) \qquad [1]$$

where $f(x, y)$ maps the original five fixed points onto each segment. For the first line segment (0,0 to 1/3,0) the function f has the form

$$f(x, y) = \underbrace{\begin{bmatrix} 1/3 & 0 \\ 0 & 1/3 \end{bmatrix}}_{A} \begin{pmatrix} x \\ y \end{pmatrix} + \underbrace{\begin{pmatrix} 0 \\ 0 \end{pmatrix}}_{t} \qquad [2]$$

where A is the shrinkage matrix and t is the translation matrix (Barnsley, 1988).

Fractal scaling may also involve translation scaling such as that needed for the line segment (2/3,0) to (1,0). In this case, the transformation function $f(x, y)$ is of the form

$$f_{DE}(x, y) = \begin{bmatrix} 1/3 & 0 \\ 0 & 1/3 \end{bmatrix} \begin{pmatrix} x \\ y \end{pmatrix} + \begin{bmatrix} 2/3 \\ 0 \end{bmatrix} \qquad [3]$$

The translation is needed to shift the fixed points to the right. The two remaining functions (for the top "hat" of the curve) also require rotation; segment BC needing to be rotated +60 degrees whereas segment

CE needs to be rotated −60 degrees. The mapping functions required for these transformation are

$$f_{BC}(x, y) = \underbrace{\begin{bmatrix} 1/2 & -\sqrt{3}/2 \\ \sqrt{3}/2 & 1/2 \end{bmatrix}}_{R} \underbrace{\begin{bmatrix} 1/3 & 0 \\ 0 & 1/3 \end{bmatrix}}_{A} \begin{pmatrix} x \\ y \end{pmatrix} + \underbrace{\begin{pmatrix} 1/3 \\ 0 \end{pmatrix}}_{t} \quad [4]$$

and

$$f_{CD}(x, y) = \overbrace{\begin{bmatrix} 1/2 & \sqrt{3}/2 \\ -\sqrt{3}/2 & 1/2 \end{bmatrix}}^{R} \overbrace{\begin{bmatrix} 1/3 & 0 \\ 0 & 1/3 \end{bmatrix}}^{A} \begin{pmatrix} x \\ y \end{pmatrix} + \overbrace{\begin{pmatrix} 1/2 \\ \sqrt{3}/6 \end{pmatrix}}^{t} \quad [5]$$

where R now represents the rotation matrix.

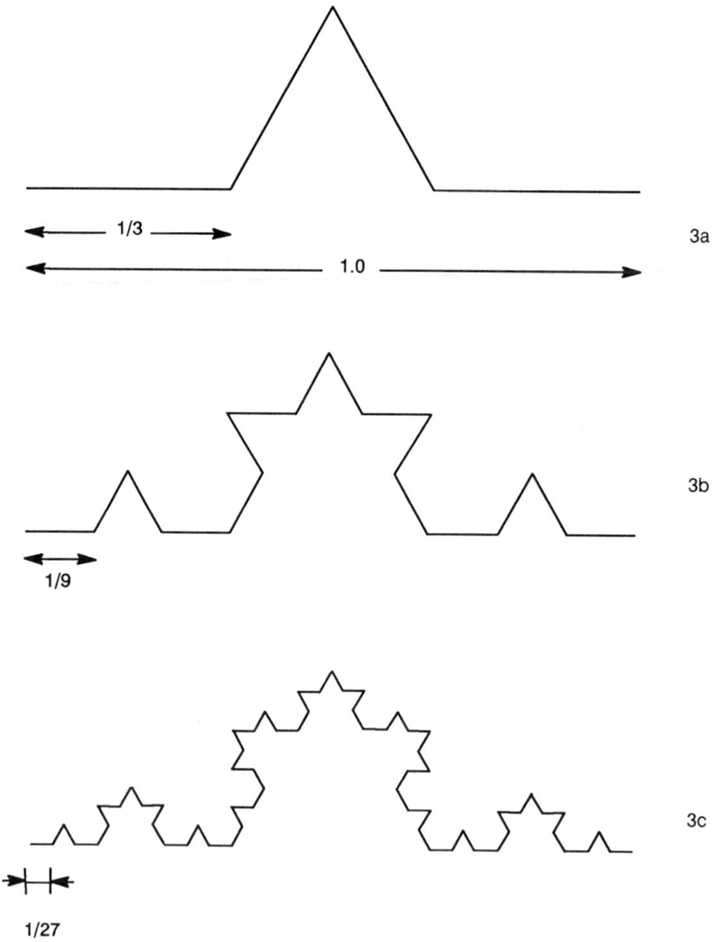

Fig. 8–3. The Koch curve carried to: (*a*) one level of recursion; (*b*) two levels; (*c*) three levels of recursion.

In each of these transformations, the A matrix remains unchanged representing the shrinkage of the original initiator. Intuitively, we can easily see this in Fig. 8–3b–c that the construction algorithm simply replaces each line segment with a scaled down (scaled in this case by a factor of 1/3) version of the initiator. Equations [2–5] simply give the algebraic rules to produce the figure.

If we complete the iteration to infinity (or a large number), we would quickly exhaust the ability of our printer to see the additional detail. If we were to magnify any portion of the curve, we would see finer and finer versions of the initiator. On the other hand, if we were to stop the algorithm after 10 iterations, the magnified curve would (near a magnification of 59 000 or $1/3^{10}$) look like a series of straight line segments. Further magnification would yield no new detail and we would say that the Koch curve showed scale invariance only over a finite range of scales. Figures 8–4a–b show this phenomenon more clearly with a random trace. In Fig. 8–4a, at a small scale of magnification, the trace appears very rough and irregular. As a portion of the trace is magnified, it is easy to tell that the rugged trace is in actuality made up of a series of small straight line segments. Figure 8–4b, however, shows the same irregular trace yet at closer inspection (higher magnification) additional detail becomes visible. At the largest magnification, it is impossible to tell the scale in viewing unless the reader is provided with a frame of reference. In these examples, Fig. 8–4a is not invariant under scale transformations, whereas Fig. 8–4b has been constructed to be scale invariant over the scales of observation.

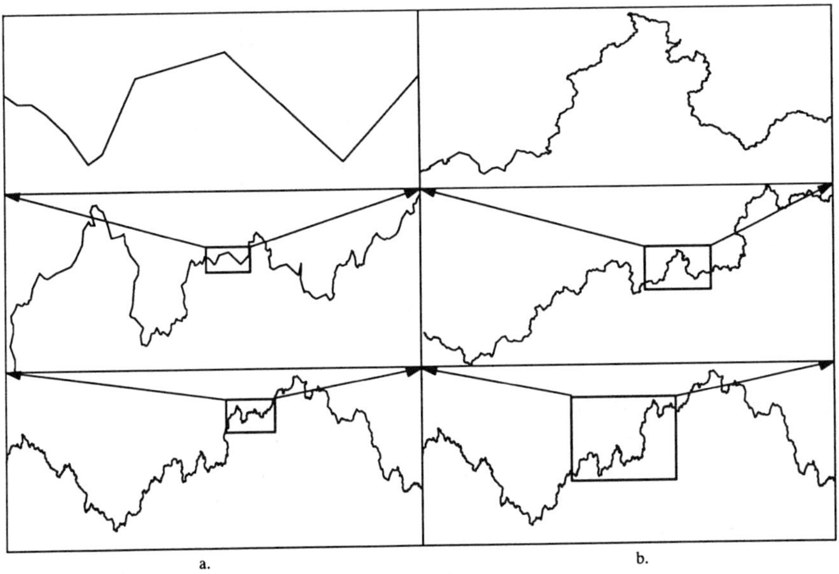

Fig. 8–4. (a) Nonfractal scaling; (b) fractal scaling.

THE CONCEPT OF THE FRACTAL DIMENSION

While at first counterintuitive, the concepts of fractional or fractal dimensions can easily be understood if thought of in terms of scale-invariant transforms and a revisit of the Koch curve of Fig. 8–3a–c. If we measure the length of the line shown in Fig. 8–3a using a measuring unit of length 1/3, the total length is obviously

$$L(\varepsilon) = N\varepsilon = 4/3 \qquad [6]$$

where N is the number of steps of size ε and $L(\varepsilon)$ is the length. If we were to measure the curve shown in Fig. 8–3b using the same measuring unit of 1/3, we would once again measure 4/3 because the measure would not be able to follow the more rugged outline any closer.

The length of Fig. 8–3b as measured in length units of 1/9 is simply

$$L(1/9) = (4 \times 4)(1/3 \times 1/3) = 1.778 \qquad [7]$$

Once again, if we used a measuring stick of length 1/9 to measure Fig. 8–3c, we would measure the same length. From these examples, it is obvious that the length $L(\varepsilon)$ is given by

$$L(\varepsilon) = (4/3)^n \qquad [8]$$

where n represents the number of scale transforms or iteration ($n = 1$ for the first iteration). The measuring unit ε is clearly

$$\varepsilon = (1/3)^n \quad \text{or} \quad n = \frac{-\log(\varepsilon)}{\log(3)} \qquad [9]$$

Substituting n back into Eq. [8] and taking anti-logs yields

$$L(\varepsilon) = \varepsilon^{[(\log 4/\log 3)-1]} \qquad [10]$$

If we let $D = \log(4)/\log(3)$ we get the now traditional fractal scaling equation

$$L(\varepsilon) = \varepsilon^{D-1} \qquad [11]$$

where D is called the fractal dimension. The numerator of D (log4) represents the number of line segments to be scaled, whereas the denominator is derived from the scale factor 1/3. Equation [11] shows the characteristics of fractal or scale-invariant transforms. As ε is decreased, the estimated length will grow provided D is greater than unity. The fractal dimension will always be greater than 1 when n exceeds the inverse of the scaling factor. In qualitative terms, a process that shows a great deal of variation will have a characteristically large fractal dimension. This is clear because in general

$$D = \frac{\log (N)}{\log (1/r)} \quad [12]$$

where r is the scaling factor. A coastline with large variations at all scales will have a high ratio of new features (N) to the scale factor (r).

An important physical aspect of fractal scaling is the measurement of length. In traditional Euclidean geometry, the notion of length is quantifiable, easily measured, and independent of measuring scale. Such is not the case with fractals. We have already shown from the simple Koch curve that the length is a function of measurement scale. For any fractal trace, the most general length relation is given by

$$L_f = \varepsilon^{1-D} H_d \quad [13]$$

where H_d is defined as the Hausdorf measure and acts as a constant of proportionality. We can express this constant of proportionality in a more useful manner for our subsequent purposes. In Fig. 8–5, we show a fractal line that can be thought of as the trace of a hypothetical particle moving through a soil pore. We impose an x–y coordinate system on it so that it intercepts the starting and ending points of the trace. Let L_s represent the straight line distance between the starting and ending points of the path. The actual length of the fractal line, or path, depends on the value of ε chosen to evaluate it. If this fractal line represents the path of a particle in a porous medium however, we recognize there is a lower limit to fractal behavior. The result is that if we zoom in on the path in Fig. 8–5 to a high enough magnification, the path will become rectifiable, that is it will have a finite length, which is shown in the blown-up portion of Fig. 8–5. This lower limit to fractal behavior, which we call ε_c, must exist, otherwise the particle will have an infinite travel distance along the fractal path, which

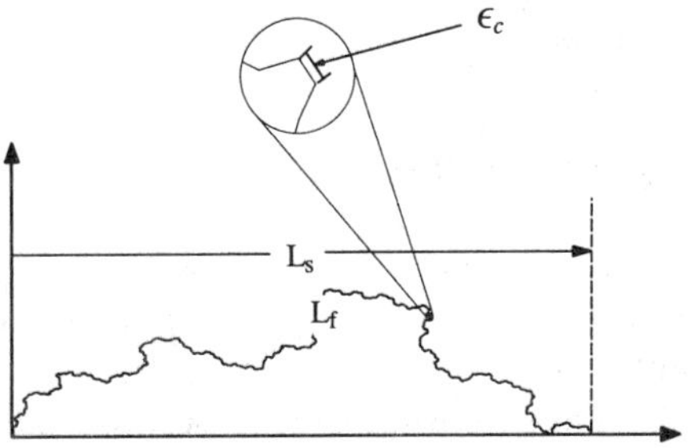

Fig. 8–5. Definition sketch for the fractal scaling relationships.

is physically unreasonable. If we set our ruler size $\varepsilon = L_s$, then it only takes one ruler length to cover the line, hence $N = 1$, and

$$H_d = L_s^D \qquad [14]$$

Substituting this relationship into Eq. [13], we have

$$L_f = \varepsilon^{1-D} L_s^D \qquad [15]$$

Equation [15] allows us to calculate the distance along a fractal path for a given ε when we know the fractal dimension and the distance along the x axis (straight-line distance) of the path. If we replace L with x in Eq. [15], and substitute ε_c for ε, we have

$$x_f = \varepsilon_c^{1-D} x_s^D \qquad [16]$$

Equation [16] is the basic fractal scaling relationship. It tells us the actual distance traveled by a particle moving along a fractal path, based on the straight-line distance between two points on a cartesian axis that intersects the fractal path. Important assumptions made are that the fractal dimension, D, is constant, and that ε_c has a meaningful average. Another important limitation is that Eq. [16] must only be used for travel distances $x_s \gg \varepsilon_c$. Scaling relationships for distances $\leq \varepsilon_c$ will be nonfractal (i.e., linear) and there may be a transitional zone of several times ε_c before full fractal behavior is developed. Assuming that ε_c is on the order of, or smaller than the pore scale, this limitation will be unimportant for field-scale transport problems. A number of relationships follow from Eq. [16] rather simply. If we divide both sides of Eq. [16] by x_s, we obtain

$$\frac{x_f}{x_s} = \varepsilon_c^{1-D} x_s^{D-1} \qquad [17]$$

Equation [17] predicts that the actual distance along a fractal path (x_f) will increase faster than the distance as measured on a cartesian axis (x_s). This equation also shows that calculations of tortuosity in a fractal medium would be scale-dependent, because the tortuosity is a measure of straight-line distance traveled compared with actual travel distance along the tortuous path that a particle might take.

EXTENSIONS TO NATURAL SYSTEMS

In the past few years, the reporting of fractal applications to real systems has grown almost as fast as Eq. [11]. Areas that seem to have benefitted significantly are those of nonlinear systems, turbulence, small-particle morphology, and geomorphology. Because this only represents a few disciplines, we suggest that the reader peruse the text of Feder (1988), which contains a wealth of fractal applications to real systems. In the following sections, we describe evidence suggesting the validity of fractal scaling in soil and aquifer systems as well as provide a brief example of fractals in soil water retention.

We often do not think of soils or aquifers as being self-similar as shown in the Koch curve. Instead, two schools of thought have developed to treat the disordered systems we call natural porous media. Both of these techniques rely upon the validity of the averaging process to move from one scale of variability to another. The more traditional concept of averaging in porous media is the idea of a representative elementary volume (REV) developed by Bear (1972). In the REV, the porous medium (soil or aquifer) is considered to be a random collection of voids and solids. Below a certain length (or volume) scale, a sample of the medium will show a wide variation as a function of the sample's location. Above a critical size of sample, the intrinsic property of interest, for example porosity, will be constant. At this scale, the intrinsic property, which represents an average, is stationary in space. The value of the property may then be assigned to the centroid of the REV. The REV approach views the porous medium as being composed of building blocks of homogeneous matrial above some critical scale. The intrinsic property assigned to these blocks may be a function of space, however in a deterministic manner.

Recent theories of stochastic or geostatistical origin propose a slightly different viewpoint. While embracing the REV concepts at some small scale [perhaps at the core scale (cm)], statistical correlation of the intrinsic property over a much wider scale are allowed (see, for example, Journel & Huijbregts, 1978). Such correlations are limited in scale, however, and statistical independence is assumed at some large, field scale. These approaches have gained wide acceptance in soils and geology, because it is intuitively obvious that many intrinsic properties will be correlated in space. Such spatial covariances, if unaccounted for, would lead to errors in the averaging process.

The basic processes generating spatial correlation, however, are often ignored in geostatistical approaches. Fractal scaling offers another view of spatially correlated data, which may account for the underlying physics. The approach suggests that over a wide range of scales, self-similar scaling can be used to predict behavior. The concept also predicts that intrinsic properties, such as the length as shown in the Koch curve, are a function of the scale of measurement. Such an approach is not consistent with the REV or geostatistical treatment and introduces some new problems in the formulation of balance or governing equations relying on averaged values of intrinsic properties. To illustrate the concepts of fractal scaling in real soil systems, consider the photographs of a "typical" soil pit in Fig. 8–6a and the geologic outcrop shown in Fig. 8–6b. The soil cross-section in Fig. 8–6a is contained within the geologic outcrop of Fig. 8–6b. Without the added visual clues of Fig. 8–6b (one of the authors), it would be difficult, if not impossible, to tell that Fig. 8–6b was taken at a much larger scale. In this case, the self-similarity of the patterns of deposition are stochastic rather than deterministic as shown in the Koch curve, or in Fig. 8–2.

This simple example illustrates the potential for fractal scaling to improve our description of heterogeneity of field soils and aquifers. It is clear that the processes that form our field soils act over a wide range of scales,

Fig. 8–6. (*a*) Close-up view of soil profile; (*b*) geologic outcrop (larger view of same profile).

from the microbiological to the continental weather systems and, as a result, correlation of intrinsic properties may be present over a much wider range of scale than we traditionally believe. These correlations cannot be ignored and, in many cases, may play a crucial role in the understanding of transport mechanisms in heterogeneous soils and aquifers. In the following example, we suggest a simple application of fractal scaling to pore and laboratory scale processes.

WATER RETENTION AND POWER-LAW SCALING

In Tyler and Wheatcraft (1990), we present a simple analysis of the power-law model of water retention developed by Brooks and Corey (1964) and Campbell (1974). In this analysis, we shall start with a very simplistic porous medium as show in Fig. 8–7. The medium is based on a simple fractal; the Sierpinski carpet. The Sierpinski carpet is generated by starting with an initial square of side length a and removing one or more subsquares of size a/b_1. This is done to the ith recursion level for increasing values of b where $b = b_1^i$. The result is a carpet everywhere filled with holes. The carpet shown in Fig. 8–7 has only been taken to two levels of recursion to more clearly show the porous structure. The carpet displays fractal scaling,

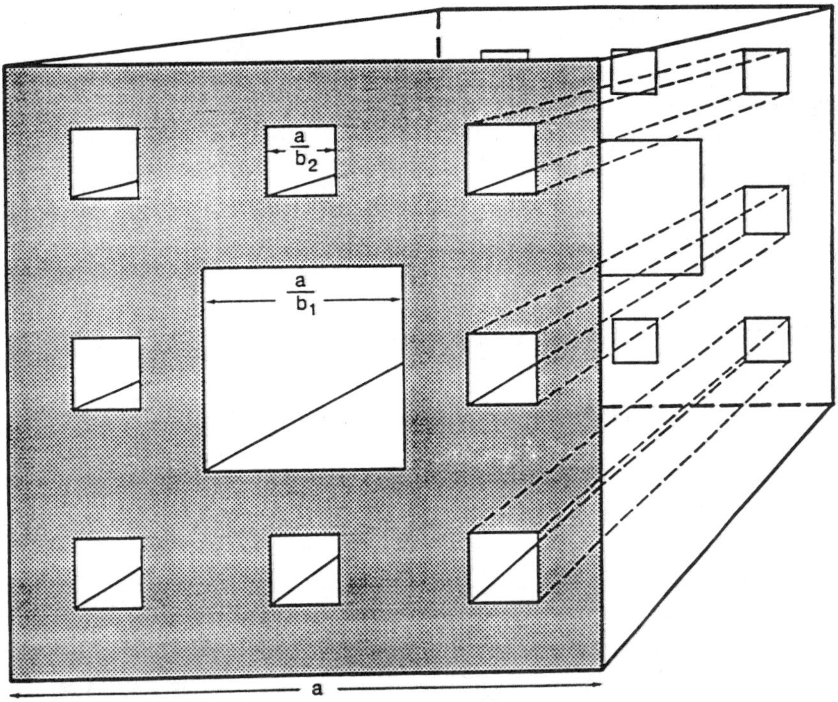

Fig. 8–7. Simulated soil using Sierpinski carpet as a conceptual model of pore structure.

in that it appears similar at any scale view, provided the scale is less than a, the initial unit length. If we assume that the open area represents the cross-section of capillary tubes, we have a very simple model of a soil's pore structure. As we carry the recursion level toward infinity, the pore radius at each recursion level becomes progressively smaller. After an infinite number of recursions, the carpet is made up of a distribution of a few large holes (pores) and a great number of small ones. In this simple model, we are only concerned with pore-number and pore-size distribution and, as such, do not treat the solid phase of the soil.

The porosity, $\phi(b)$, of the carpet, after any arbitrary recursion level is given by

$$\phi(b) = 1 - b^{D-2} \qquad [18]$$

where b is inversely related to the smallest pore size and D is the fractal dimension of the carpet. The fractal dimension of the carpet sets the recursion algorithm and is given by

$$D = \frac{\log\,(b_1^2 - l_1)}{\log\,(b_1)} \qquad [19]$$

where $1/b_1$ represents the characteristic size of the largest pore and l_1 represents the number of pores of size $1/b_1$ removed at the first recursion level.

As b gets larger, i.e., large iterations, the total porosity of the carpet approaches unity, as can be seen in Eq. [18]. The volumetric water content, $\theta(b)$, held in the pores of size $1/b$ or smaller, from Eq. [18] is simply

$$\theta(b) = \phi(b = \infty) - \phi(b) \qquad [20a]$$

or

$$\theta(b) = b^{D-2} \qquad [20b]$$

From the capillary rise equation, we know that

$$\psi \; a \; \frac{1}{R} \; a \; b \qquad [21]$$

where R represents the equivalent pore radius.

Substituting Eq. [21] into Eq. [20b] directly yields a power-law relation of water retention

$$\theta(\psi) \; a \; \psi^{D-2} \qquad [22]$$

We can normalize Eq. [22] in terms of saturated water content, θ_s, to get the familiar Brooks and Corey (1964) or Campbell (1974) form

$$\frac{\theta}{\theta_s} = \left(\frac{\psi}{\psi_a}\right)^{D-2} \qquad [23]$$

where ψ_a is the air entry pressure.

The advantage of Eq. [23] is that the exponent $(D-2)$ is not a curve-fitting parameter, but rather is strictly defined and controls pore-number and pore-size distribution. Various examples of values of D are given in Tyler and Wheatcraft (1990) along with their calculated retention functions. These data show that D is inversely proportional to soil texture; with larger values of D corresponding to finer textured soils. This behavior is consistent with the empirically derived values of the power-law exponent of Brooks and Corey, b_c, or Campbell's λ. This simple model suggests that the nature of the pore structure may be directly derived from the power-law exponent of retention data and is fractal in nature. This direct interpretation is advantageous over simple curve fitting and may be very useful in prediction of the hydraulic conductivity of the soil.

CONCLUSION

The concept of fractal scaling offers a new viewpoint on quantifying the spatial variability of soil and aquifer properties. Fractal scaling suggests that soil properties will be a function of the scale of measurement and that traditional concepts of stationarity and averaging may not capture the total nature of the heterogeneity. It is clear that soil and aquifer-forming processes act over a wide range of scales, from millimeters to kilometers, and it is reasonable to assume that these processes will produce correlation in some properties over such a range of scales. Based on the successful use of fractals to describe a wide spectrum of physical processes and properties, it appears that it will be a valuable descriptor of the nature of soil spatial variability. In this chapter, we have presented a brief description of fractal scaling based on the notion of scale invariant transformations. These techniques, developed in detail by Barnsley (1988), offer the soil scientist the analytical tools necessary to investigate for fractal scaling in soil properties. It is our hope that this work will stimulate work in this area and shed new light on the quantification of soil spatial variability.

ACKNOWLEDGMENT

This work was funded by the Nevada Agency for Nuclear Projects/ Nuclear Waste Projects Office under Dep. of Energy Grant no. DE-FG08-85-NV10461. The opinions expressed in this chapter do not necessarily represent those of the State of Nevada or the U.S. Dep. of Energy.

REFERENCES

Barnsley, M. 1988. Fractals everywhere. Academic Press, New York.

Bear, J. 1972. Dynamics of fluids in porous media. Elsevier Publ., New York.

Brooks, R.H., and A.T. Corey, 1964. Hydraulic properties of porous media. Hydrol. Pap. 3. Colorado State Univ., Fort Collins, CO.

Campbell, G.S. 1974. A simple method for determining unsaturated hydraulic conductivity from moisture retention data. Soil Sci. 117:311–314.

Feder, J. 1988. Fractals. Plenum Press, New York.

Journel, A.G., and Ch.J. Huijbregts. 1978. Mining geostatistics. Academic Press, New York.

Mandelbrot, B.B. 1983. The fractal geometry of nature. W.H. Freeman, San Francisco.

Miller, E.E. 1980. Similitude and scaling of soil water phenomena. p. 300–318. *In* D. Hillel (ed.) Applications of soil physics. Academic Press, New York.

Shouse, P., J. Scisson, G. de Rooij, J. Jobes, and M.Th. van Genuchten. 1989. Application of fixed-gradient methods for estimating soil hydraulic conductivity. *In* M.Th. van Genuchten et al. (ed.) Proceedings of International Workshop on Indirect Methods for Estimating the Hydraulic Properties of Unsaturated Soils, Riverside, CA. 11–13 October. Univ. of California, Riverside, CA.

Tyler, S.W., and S.W. Wheatcraft. 1990. Fractal processes in soil water retention. Water Resour. Res. 26(5):1047–1054.